The Organic Chem Lab Survival Manual
Survival Manual
A Student's Guide to Techniques

The Organic Chem Lab Survival Manual:
A Student's Guide to Techniques

James W. Zubrick
Walker Laboratory
Rensselaer Polytechnic Institute
Troy, New York

John Wiley & Sons
New York • Chichester • Brisbane • Toronto • Singapore

Library of Congress Cataloging in Publication Data:
Zubrick, James W.
    The organic chem lab survival manual.

    Includes index.
    1. Chemistry, Organic—Laboratory manuals.   I. Title.
QD261.Z83   1984   547'.0078   83-21808
ISBN 0-471-87131-1

Printed in the United States of America

10 9 8 7 6 5 4 3 2 1

Forewords

Seldom does one have the opportunity to read and use a textbook that is completely useful, one that does not need substitutions and deletions. Zubrick's book is this type of resource for undergraduate organic students and their laboratory instructors and professors. I must heartily recommend this book to any student taking the first laboratory course in organic chemistry.

The Organic Chem Lab Survival Manual is filled with explanations of necessary techniques in much the same way that advanced techniques have been presented in books by Wiberg, Lowenthal, Newman, and Gordon and Ford. In larger universities, *The Survival Manual* is a valuable supplement to most laboratory manuals. It provides explanations that many graduate teaching assistants do not take time to give to their classes. Most teaching assistants of my acquaintance appreciate Zubrick's book because it supports their discussions during recitations (when each student has a personal copy), and it refreshes their memories of good techniques they learned and must pass on to a new generation of undergraduates.

The book is addressed to the undergraduate student audience. The informal tone appeals to most laboratory students. The illustrations are delightful. The use of different type fonts is effective for emphasis. Also, Zubrick always explains *why* the particular sequence of operations is necessary, as well as *how* to manipulate and support the apparatus and substances. This is a definite strength.

This book is an evolutionary product: over the span of a decade, professors at major universities and liberal arts colleges have made suggestions for minor changes and improvements. I count myself fortunate to have used the forerunners, which have been privately published since 1973.

A large quantity of useful information has been collected, well organized, and presented with great care. This book is the handiwork of a master teacher.

C.W. Schimelpfenig
Dallas Baptist College
Dallas, Texas

The Organic Chem Lab Survival Manual is a book I have known about for a number of years in a variety of developmental stages. As it progressed, I watched with interest as Jim Zubrick struggled to achieve a balance between merely conveying information—what most books do— and conveying that information efficiently to its very human audience. On the one hand, Jim insisted that his book contain all the conventional scientific detail; on the other, he also insisted that a "how to" book for the organic chemistry lab need not be written in the dull and confusing prose which so often passes as the *lingua franca* of science. This book demonstrates that he has achieved both goals in admirable fashion.

In fact, *The Survival Manual* succeeds very well in following Wittgenstein's dictum that "everything that can be thought at all, can be thought clearly. Anything that can be said can be said clearly." It also follows the advice of Samuel Taylor Coleridge to avoid pedantry by using only words "suitable to the time, place, and company."

Although some few readers may take umbrage with this book because it is not, atypically, couched in the language of a typical journal article, similar people also complained when William Strunk published *Elements of Style* in 1919. For Strunk also broke with tradition in writing his book. Most other texts of the day were written in the convoluted language of the eighteenth century, and the material they contained consisted largely of lists of arcane practices, taboos, and shibboleths—all designed to turn students into eighteenth-century writers. From Strunk's point of view, such texts were less than desirable for several major reasons. First, the medicine they offered students had little to do with the communication process itself; second, it had little to do with current practice; and third, taking the medicine was so difficult that the cure created more distress in the patients than did the disease itself.

Jim Zubrick proves in this book that he understands, as did Strunk, that learning reaches its greatest efficiency in situations where only that information is presented which is directly related to completing a specific task. In an environment fraught with hazards, efficiency of this sort becomes even more necessary.

The Survival Manual is an excellent book because it speaks to its audience's needs. Always direct—if sometimes slightly irreverent—the book says clearly what many other books only manage to say with reverent indirection. It never forgets that time is short or that the learning curve rises very slowly at first. The prose is straightforward, easy to understand, and is well supported by plentiful illustrations keyed to the text. It is also technically accurate and technically complete, but it always explains matters of laboratory technology in a way designed to make them easily understandable to students in a functional context.

All of these characteristics related to communication efficiency will naturally make the laboratories in which the book is used safer labs; the improved understanding they provide serves as natural enhancement to the book's emphatic and detailed approach to safety in the laboratory.

Most important, however, all the elements of *The Survival Manual* come together in focusing on the importance of task accomplishment in a way which demonstrates the author's awareness that communication which does not efficiently meet the needs of its audience is little more than *pedantry* unsuitable to the time, place, and company.

David. L. Carson
Director,
The Master of Science
Programs in Technical Writing
and Communication
Rensselaer Polytechnic Institute

Preface

Describe, for the tenth time, an instrument not covered in the laboratory book, and you write a procedure. Explain, again and again, operations that are in the book, and you get a set of notes. When these produce questions, you revise until the students, not you, finally have it right. If you believe that writing is solidified speech—with the same pauses, the same cadences—then a style is set. And if you can still laugh, you write this book.

This book presents the basic techniques in the organic chemistry laboratory with the emphasis of doing the work correctly the first time. To this end, examples of what can go wrong are presented with admonishments, often bordering on the outrageous, to forestall the most common of errors. This is done in the belief that it is much more difficult to get into impossible experimental troubles once the student has been warned of the merely improbable ones. Complicated operations, such as distillation and extraction, are dealt with in a straightforward fashion, both in the explanations and in the sequential procedures.

The same can be said for the sections concerning the instrumental techniques of GC, IR, NMR, and HPLC. The chromatographic techniques of GC and HPLC are presented as they relate to thin-layer and column chromatography. The spectroscopic techniques depend less on laboratory manipulation and so are presented in terms of similarities to the electronic instrumentation of GC and HPLC techniques (dual detectors, UV detection in HPLC, etc.). For all techniques, the emphasis is on correct sample preparation and correct instrument operation.

Many people deserve credit for their assistance in producing this textbook. It has been more than a few years since this book was first written, and a list of acknowledgments would approach the size of a small telephone directory—there are too many good people to thank directly.

For those who encouraged, helped, and constructively criticized, thanks for making a better book that students enjoy reading and learning from.

I'd like to thank the hundreds of students who put up with my ravings, rantings, put-ons, and put-downs, and thus taught me what it was they needed to know, to survive organic chemistry laboratory.

A special thanks to Dr. C.W. Schimelpfenig, for encouragement over many years when there was none, and whose comments grace these pages; Dr. D.L. Carson, whose comments also appear, for his useful criticism concerning the presentation; Drs. R.A. Bailey, S.C. Bunce, and H.B. Hollinger for their constant support and suggestions; Dr. Mark B. Freilich, whose viewpoint as an inorganic chemist proved valuable during the review of the manuscript; and Dr. Sam Johnson, who helped enormously with the early stages of the text processing. I also thank Christopher J. Kemper and Keith Miller for their valuable comments on the instrumental sections of the book.

Finally, I'd like to thank Clifford W. Mills, my patron saint at John Wiley & Sons, without whose help none of this would be possible, and Andrew E. Ford, Jr., vice president, for a very interesting start along this tortured path to publication.

Some Notes on Style

It is common to find instructors railing against poor usage and complaining that their students cannot do as much as to write one clear, uncomplicated, communicative English sentence. Rightly so. Yet I am astonished that the same people feel comfortable with the long and awkward passive voice, the pompous "we" and the clumsy "one," and that damnable "the student," to whom exercises are left as proofs. These constructions, which appear in virtually all scientific texts, do *not* produce clear, uncomplicated, communicative English sentences. And students do learn to write, in part, by following example.

I do not go out of my way to boldly split infinitives, nor do I actively seek prepositions to end sentences with. Yet by these constructions alone, I may be viewed by some as aiding the decline in students' ability to communicate.

E.B. White, in the second edition of *The Elements of Style* (Macmillan, New York, 1972, p. 70), writes

> Years ago, students were warned not to end a sentence with a preposition; time, of course, has softened that rigid decree. Not only is the preposition acceptable at the end, sometimes it is more effective in that spot than anywhere else. "A claw hammer, not an axe, was the tool he murdered her with." This is preferable to "A claw hammer, not an ax, was the tool with which he murdered her."
>
> Some infinitives seem to improve on being split, just as a stick of round stovewood does. "I cannot bring myself to really like the fellow." The sentence is relaxed, the meaning is clear, the violation is harmless and scarcely perceptible. Put the other way, the sentence becomes stiff, needlessly formal. A matter of ear.

We should all write as poorly as White.

The elimination of sexist language is also a concern. Unless I know the person being addressed or described, I use the male pronouns exclusively. Many do not believe women are good enough for these and insist on inferior expressions. The "he/she" construction has been suggested. Typographically, however, it looks separate but equal and that attitude has led to even more serious problems.

With the aid of William Strunk and E.B. White in *The Elements of Style* and that of William Zinsser in *On Writing Well*, Rudolph Flesch in *The ABC of Style*, and D.L. Carson, whose comments appear in this book, I have tried to follow some principles of technical communication lately ignored in scientific texts: use the first person, put yourself in the reader's place, and, the best for last, use the active voice and a personal subject.

The following product names belong to the respective manufacturers. Registered trademarks are indicated here, as appropriate; in the text, the symbol is omitted.

Corning®	Corning Glass Works, Corning, New York
Drierite®	W. A. Hammond Drierite Company, Xenia, Ohio
Fisher–Johns®	Fisher Scientific Company, Pittsburgh, Pennsylvania
Luer-Lok®	Becton, Dickinson and Company, Rutherford, New Jersey
Mel-Temp®	Laboratory Devices, Cambridge, Massachusetts
Millipore®	Millipore Corporation, Bedford, Massachusetts
Swagelok®	Crawford Fitting Company, Solon, Ohio
Teflon®	E.I. DuPont de Nemours & Company, Wilmington, Delaware
Variac®	General Radio Company, Concord, Massachusetts

Contents

The Organic Chem Lab
Survival Manual
A Student's Guide to Techniques

Safety First, Last, and Always

The organic chemistry laboratory is potentially one of the most dangerous of undergraduate laboratories. That is why you must have a set of safety guidelines. It is a very good idea to pay close attention to these rules, for one very good reason:

The penalties are only too real.

Disobeying safety rules is not at all like flouting many other rules. *You can get seriously hurt.* No appeal. No bargaining for another 12 points so you can get into medical school. Perhaps as a patient, but certainly not as a student. So, go ahead. Ignore these guidelines. But remember,

You have been warned!

1. ***Blue Cross or Blue Shield?*** Find out how you would get to medical help, if you needed it. Sometimes during a summer session, the school infirmary is closed and you would have to be transported to the nearest hospital.

2. ***Wear your goggles.*** Eye injuries are extremely serious, but they can be mitigated or prevented if you keep your goggles on *at all times.* There are several types of eye protection available, some acceptable, some not, according to local, state, and federal laws. I like the clear plastic jobbers that leave an unbroken red line on your face when you remove them. Sure, they fog up a bit, but the protection is superb. Also, think about getting chemicals or chemical fumes trapped under your contact lenses. Then don't wear them to lab. Ever.

3. ***Touch not thyself.*** Not a Biblical injunction, but a bit of advice. You may have just gotten chemicals on your hands, in a concentration that is not noticeable. Sure enough, up go the goggles for an eye wipe with the fingers. Enough said.

4. ***There is no "away."*** Getting rid of chemicals is a very big problem. Now there are some laws to stop you from poisoning someone else when you throw chemicals from here. However, the rules were designed for industrial situations, where there are hundreds of gallons of waste having the same composition. In a semester of organic lab there will be much smaller amounts of different materials. It is not practical to provide waste containers for every waste material that will be generated. If you don't see the waste can you need, ask your instructor. When in doubt, *ask.*

1

5. ***Bring a friend.*** If you have a serious accident when you are all by yourself, you might not be able to get to help before you die. Don't work alone; don't work at unauthorized times.

6. ***Don't fool around.*** Chemistry is serious business. Don't be careless or clown around in lab. You can hurt yourself or other people. Try not to be somber about it; just serious.

7. ***Drive defensively.*** Work in the lab as if someone else were going to have an accident that might affect you. Keep the goggles on because *someone else* is going to point a loaded, boiling test tube at you. *Someone else* is going to spill hot, concentrated acid on your body. Get the idea?

8. ***Eating, drinking, or smoking in lab.*** Are you kidding? Eat in a chem lab?? Drink in a chem lab??? Smoke, and blow yourself up????

9. ***Keep it clean.*** Work neatly. You don't have to make a fetish out of it, but try *neat*. Clean up spills. Turn off burners or water or electrical equipment when not in use.

10. ***Where it's at.*** Learn the location and proper use of the fire extinguishers, fire blankets, safety showers, and eyewashes.

11. ***Make the best-dressed list.*** No open-toed shoes, sandals, or canvas-covered footwear. No loose-fitting cuffs on pants or shirts. Keep the midsection covered. Tie back that long hair. And a small investment in a lab coat can pay off, projecting that extra professional touch. It gives a lot of protection too.

12. ***Hot under the collar.*** Many times you'll be asked or told to heat something. Don't automatically go for the Bunsen burner. That way lies *fire*. Usually—

No flames!

Try a hot plate, try a heating mantle (see Chapter 13, "Sources of Heat"). But try to stay away from flames. Most of the fires I've had to put out started when some bozo decided to heat up a flammable solvent in an open beaker. Sure, there are times when you'll HAVE to use a flame. But use it away from all flammables. In a hood (Fig. 1). With permission of your instructor.

13. ***Work in the hood.*** A hood is a specially constructed place to work and has, at the least, a powered vent to suck all the noxious fumes outside. There's also a safety glass or plastic panel you can slide down

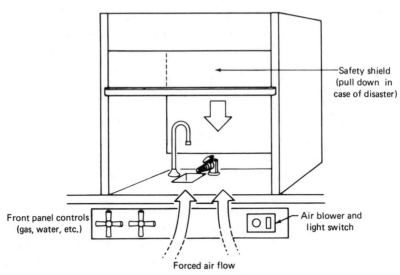

Fig. 1 A typical hood.

as protection from exploding apparatus (Fig. 1). If it is at all possible, treat every chemical (even solids) as if toxic or bad-smelling fumes came off it, and carry out as many of the operations in the organic lab as possible *inside a hood,* unless told otherwise.

These are a few of the safety guidelines for an organic chemistry laboratory. You may have others particular to your own situation.

ACCIDENTS WILL *NOT* HAPPEN

That's an attitude you might hold while working in the laboratory. You are NOT going to do anything, or get anything done to you, that will require medical attention. If you do get cut, and the cut is not serious, wash the area with water. If there's serious bleeding, apply direct pressure with a clean, preferably sterile, dressing. For a minor burn, let cold water run over the burned area. For chemical burns to the eyes or skin, flush the area with lots of water. In every case, get to see a physician if at all possible.

If you have an accident, *tell your instructor immediately. Get help!* This is no time to worry about your grade in lab. If you put grades ahead of your personal safety, be sure to see a psychiatrist after the internist finishes.

Read This
This
Next!!

2

Unlike many distinguished texts, this book is not intended to be, nor does it have any pretensions of being, the end-all and be-all of all lab texts. It is just something to get you working efficiently in the lab. The style is designed to hold your attention. I want you to enjoy reading these procedures and hope that you'll be entertained or shocked into remembering what has to be done.

Read the chapters on laboratory technique immediately. All of them. If you don't, it is highly likely that you'll get lost. The payoff will be improved understanding, and for those of you less highly motivated,

Higher grades!

(unless you're a glutton for punishment and LIKE spending lots of wasted hours in lab). You might be able to hold off reading the instrumentation chapters until later in the course of the lab; but there is no way you can work in the lab without the general laboratory techniques.

Since this text comes from my personal, highly opinionated lecture notes, I suggest you read them aloud to your friends, *hollering all passages in italics, while standing on a chair.*

Day One: The Initiation

3

After today you will be glad to return to the quiet serenity of a boiler factory and give your nerves a well-deserved rest. Here is a glimpse of the fun and frolic in store for you.

You show up in a room and wait for your instructor. The schedule may list him as a "Mr. Staff," who is teaching all the other sections as well. That's the university for you—the personal touch.

Here is the first disappointment. It seems that Mr. Staff has been replaced by an ordinary living, breathing human being. Look carefully upon his countenance, memorize his name, and find his office. He is the one who will decide your lab grade and influence several of your hours every week. He may be enthusiastic or apathetic or funny or boring or anything. Remember this—his word is law. If your instructor's technique is different from mine,

By all means follow your instructor!

His detailed knowledge adds to the information presented here. It will be a big help, since there are differences in equipment, philosophy, and so on. If you don't like *this* approach, remember

He is determining your grade—not I!

Soon the lab begins. You rejoice. Boredom ceases. Confusion reigns!

Keeping a Notebook

4

A **research notebook** is perhaps one of the most valuable pieces of equipment you can own. With it you can duplicate your work, find out what happened at leisure, and even figure out where you blew it. General guidelines for a notebook are:

1. The notebook must be bound permanently. No loose-leaf or even spiral-bound notebooks will do. It should have a sewn binding so that the only way pages can come out is to cut them out ($8\frac{1}{2} \times 11$ in. is preferred).
2. *Use ink! Never pencil!* Pencil will disappear with time, and so will your grade. Never erase! Just draw *one* line ~~thruogh yuor errers~~ through your errors so that they can still be seen. And never, never, never cut any pages out of the notebook!
3. Leave a few pages at the front for a table of contents.
4. Your notebook is your friend, your confidant. Tell it:
    a. What you have done. Not what it says to do in the lab book. What you, yourself, have done.
    b. Any and all observations: color changes, temperature rises, explosions, . . ., anything that occurs. Any *reasonable* explanation *why* whatever happened happened.
5. Skipping pages is in *extremely* poor taste. It is NOT done!
6. List the IMPORTANT chemicals you'll use during each reaction. You should include USEFUL **physical properties:** the name of the compound, molecular formula, molecular weight, melting point, boiling point, density, and so on. The *CRC Handbook of Chemistry and Physics,* originally published by the Chemical Rubber Company and better known as the *CRC Handbook,* is one place to get this stuff. (I now have some rather uncomplimentary things to say about these fellows in Chapter 5, "Interpreting a Handbook").

    Note the qualifier "USEFUL." If you can't use any of the information given, do without it! You look this up *before* the lab so you can tell what's staring back out of the flask at you during the course of the reaction.

Once again, if your instructor wants anything different, do it for him. The art of notebook keeping has many schools—follow the perspective of your own.

Interpreting
A
Handbook

5

You should look up information concerning any organic chemical you'll be working with so that you will know what to expect in terms of molecular weight, density, solubility, crystalline form, melting or boiling point, color, and so on. This information is kept in handbooks that should be available in the lab, if not in the library. Reading some of these is not easy, but once someone tells you what the fancy symbols mean, there shouldn't be a problem. Many of the symbols are common to all handbooks and are discussed only once, so read the entire section even if your handbook is different. There are at least four fairly popular handbooks.

1. ***CRC Handbook.*** (*CRC Handbook of Chemistry and Physics,* CRC Press, Inc., Boca Raton, Florida.) Commonly called "the CRC," as in, "Look it up in the CRC." A very popular book; a classic. Sometimes you can get the past year's edition cheaply from the publisher, but usually you have to order 10 copies or more. Even so, stay away from their 61st edition. The melting point, boiling point, and formula indexes are missing. Oh, yes. The solubilities aren't really distinguished any more (see below). Various solvents are listed without comment. They've kept this up, so don't get the newer editions.

2. ***Lange's.*** (*Lange's Handbook of Chemistry,* McGraw-Hill Book Company, New York.) A fairly well-known, but not well-used, handbook. It lists some compounds by common names, which can be very handy.

3. ***Merck Index.*** (*The Merck Index,* Merck & Company, Inc., Rahway, New Jersey.) This handbook is mostly concerned with drugs and their physiological effects, but it gives useful information concerning many chemicals. It's great for synonyms.

4. ***The Aldrich Catalog.*** (Aldrich Chemical Company, Inc., Milwaukee, Wisconsin.) Not your traditional hardbound reference handbook, but handy nonetheless. Aldrich makes many compounds, some not yet listed in the other handbooks, and often the catalog gives structures and physical constants.

CASTING ABOUT THE RUNES

You'll see abbreviations for physical properties in all handbooks, and there's usually a key or table that explains them. Here are some examples of the terms in these handbooks.

1. ***M.P. 80.*** Handbooks report only the TOP of the melting point range. You, however, should report the *entire* range. They also assume you know it's 80°C.
2. ***B.P.*** 75^{50}***.*** This indicates that the boiling point was taken at 50 torr, that is, a vacuum distillation. Without any superscript (or subscript), assume atmospheric pressure (760 torr).
3. ***Density*** 1.021^{20}***.*** This is an actual density, in grams per milliliter (g/ml) at 20°C. Another expression is the specific gravity. It's the ratio of the weight of a given volume of your liquid to that of the weight of the same volume of water. The water standard is usually taken at 4°C when the density of water is 1.000 g/ml, so the specific gravity winds up being numerically equal to the density but without any units. 1.012^{15}_{4} shows that the specific gravity was taken at 15°C and referred to (divided by) the density of water at 4°C. Other temperature ratios exist.
4. ***Refractive index*** n_D 1.021^{20}***.*** Obtained using the yellow light from a sodium lamp (the D line) at 20°C.
5. ***Crystalline form nd (alc).*** Needlelike crystals when recrystallized from alcohol. It's nice having the recrystallization solvent listed, in case no one told you. Some other abbreviations include

lf.	leafs	pr.	prisms	pl.	plates
cr.	crystals	ye	yellow	pa	pale

Thus, pa ye nd (w) means that pale yellow needles are obtained when you crystallize the compound from water.

Sometimes solubilities are given for a compound as well. Some of the more popular abbreviations are

s	soluble	i	insoluble
δ	slightly soluble	∞	miscible, mixes in all proportions
h	solvent must be hot	v	very; also goes in front of s or i to add new meanings to these otherwise mundane letters

As I said, the 61st and later editions of the *CRC Handbook* list solvents for some compounds WITHOUT any comment, for example,

Name	Solubility
Benzoic acid	al, eth, ace, bz, chl

Thus, benzoic acid is soluble in alcohol, ether, acetone, benzene, and chloroform. The 60th edition, and some earlier ones, are different, for example,

Name	w	al	eth	ace	bz	Other
Benzoic acid	δ s^h	v	v	s	v^h s	CCl_4s lig δ to, chl, MeOH s

I don't like the change. The earlier listing gives you enough information to judge what solvents you can dissolve benzoic acid in.

P.S. w is water; al is alcohol, *specifically ethyl alcohol;* eth, ether, *specifically diethyl ether;* lig is ligroin, a hydrocarbon; to is toluene; MeOH means methanol; CCl_4 is carbon tetrachloride.

Jointware

6

Using **standard taper jointware** you can connect glassware without rubber stoppers, corks, or tubing. Pieces are joined by glass connections built into the apparatus (Fig. 2). They are manufactured in standard sizes, and you'll probably use $19/22.

The symbol $ means **standard taper.** The first number is the size of the joint at the widest point, in millimeters. The second number is the length of the joint, in millimeters. This is simple enough. Unfortunately, life is not all that simple, except for the mind that thought up the next devious little trick.

STOPPERS WITH ONLY ONE NUMBER

Sounds crazy, no? But with a very little imagination, and even less thought, grave problems can arise from confusing the two. Look at Fig. 3, which shows that all glass stoppers are not alike. Interchanging these two leads to **leaking joints** through which your **graded** product can escape. Also, the $19/22 stopper is much more expensive than the $19 stopper, and you may *have to pay money* to get the correct one when you check out at the end of the course. Please note the emphasis in the last two sentences. I appeal to your better nature and common sense. Take some time to check these things out.

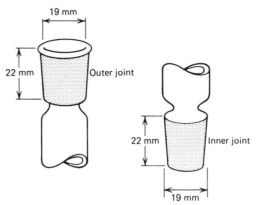

Fig. 2   Standard taper joints ($ 19/22).

Fig. 3 A ⊤ 19 nonstandard stopper in a ⊤ 19/22 standard taper joint.

As you can see from Fig. 3, that single number is the width of the stopper at its top. There is no mention of the length, and you can see that it is too short. The ⊤19 stopper *does not* fit the ⊤19/22 joint. Only the ⊤19/22 stopper can fit the ⊤19/22 joint. Single-number stoppers are commonly used with volumetric flasks. Again, they will leak or stick if you put them in a double-number joint.

With these delightful words of warning, we continue the saga of coping with ground glass jointware. Figures 4 and 5 show some of the more familiar pieces of jointware you may encounter in your travels. They may not be familiar to you now, but give it time. After a semester or so, you'll be good friends, go to reactions together, maybe take in a good synthesis. Real fun stuff!

These pieces of jointware are the more common pieces that I've seen used in the laboratory. You may or may not have *all* the pieces shown in Figs. 4 and 5. Nor will they necessarily be called EXACTLY by the names given here. The point is *find out* what each piece is, and *make sure* that it is in good condition *before* you sign your life away for it.

ANOTHER EPISODE OF *LOVE OF LABORATORY*

"And that's $28.46 you owe us for the separatory funnel."
"But it was broken when I got it!"
"Should've reported it then."
"The guy at the next bench said it was only a two-dollar powder funnel

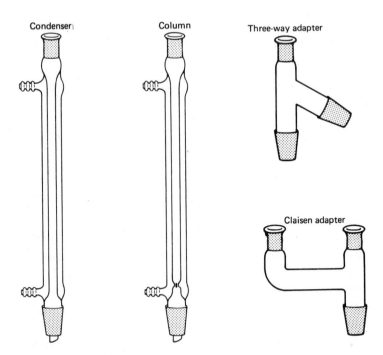

Fig. 4 Some jointware.

and not to worry and the line at the stockroom was long anyway, and . . . and . . . anyway the stem was only cracked a little . . . and it worked OK all year long. . . . Nobody said anything. . . ."
"Sorry."

Tales like these are commonplace, and ignorance is no excuse. Don't rely on expert testimony from the person at the next bench. He may be more confused than you are. And equipment that is "slightly cracked" is much like a person who is "slightly dead." There is no in-between. If you are told that you *must* work with damaged equipment because there is no replacement available, you would do well to get it in writing.

HALL OF BLUNDERS AND THINGS NOT QUITE RIGHT

Round-Bottom Flasks

Round-bottom (R.B.) jointware flasks are so round and innocent looking that you would never suspect they can turn on you in an instant.

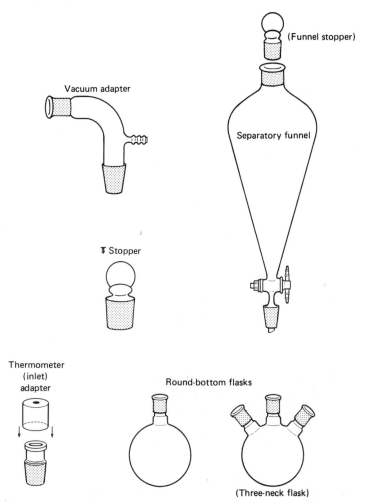

Fig. 5 More jointware.

1. **Star cracks.** A little-talked-about phenomenon that turns an ordinary R.B. flask into a potentially explosive monster. Stress, whether prolonged heating in one spot or indiscriminate trouncing on hard surfaces, can cause a flask to develop a **star crack** (Fig. 6) on its backside. Sometimes the crack is hard to see, but if overlooked, the flask can split asunder at the next lab.
2. **Heating a flask.** Since they are cold-blooded creatures, flasks show more of their unusual behavior while being heated. The behavior is

Fig. 6 R.B. flask with star crack.

usually unpleasant if certain precautions are not taken. In addition to star cracks, various states of disrepair can occur, leaving you with a benchtop to clean. Both humane and cruel heat treatment of flasks will be covered in Chapter 13, "Sources of Heat," which is on the SPCG (Society for the Prevention of Cruelty to Glassware) recommended readings list.

Columns and Condensers

A word about **distilling columns** and **condensers:**

Different!

Use the **condenser** as is for **distillation** and **reflux** (see Chapter 15, "Distillation" and Chapter 16, "Reflux"). You can use the *column with or without column packing* (bits of metal or glass or ceramic or stainless-steel sponge—whatever)! That's why the column is wider and it has **projections** at the end (Fig. 7). These projections help hold up the column packing if you use any packing at all (see Fig. 57).

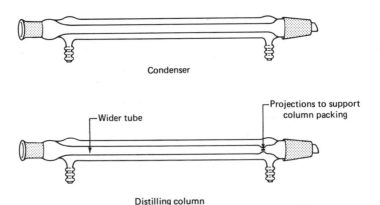

Condenser

Projections to support column packing

Wider tube

Distilling column

Fig. 7 Distilling column versus condenser.

If you jam column packing into the skinny condenser, the packing may never come out again! Using a condenser for a packed column is bad form and can lower your esteem or grade, whichever comes first.

You might use the column as a condenser.

Never use the condenser as a packed column!

THE ADAPTER WITH LOTS OF NAMES

Figure 8 shows the one place where joint and nonjoint apparatus meet. There are two parts: a rubber cap with a hole in it and a glass body. Think of the rubber cap as a rubber stopper through which you can insert thermometers, inlet adapters, drying tubes, and so on.

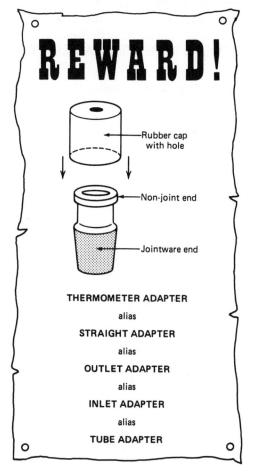

REWARD!

Rubber cap with hole

Non-joint end

Jointware end

THERMOMETER ADAPTER

alias

STRAIGHT ADAPTER

alias

OUTLET ADAPTER

alias

INLET ADAPTER

alias

TUBE ADAPTER

Fig. 8 Thermometer adapter.

6

CAUTION! Do not force. You might snap the part you're trying to insert. Handle both pieces through a cloth; lubricate (water) and then insert carefully.

The rubber cap fits over the **nonjoint** end of the glass body. The other end is a **ground glass joint** and *fits only other glass joints*.

The rubber cap should neither crumble in your hands nor need a 10-ton press to bend it. If the cap is shot, get a new one. Let's have none of these corks, rubber stoppers, chewing gum, or any other type of plain vanilla adapter you may have hiding in the drawer.

And remember: Not only thermometers but *anything* that resembles a glass tube can fit in here! This includes unlikely items such as **drying tubes** (they have an outlet tube) and even a **funnel stem** (you may have to couple the stem to a smaller glass tube if the stem is too fat).

SOCIALLY ACCEPTABLE THINGS TO DO WITH THE ADAPTER WITH LOTS OF NAMES

The imaginative arrangements shown in Fig. 9 are acceptable.

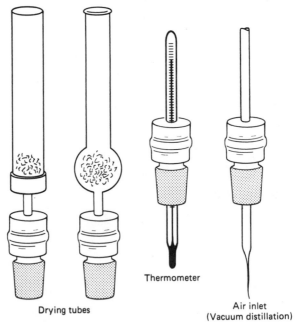

Thermometer

Drying tubes

Air inlet
(Vacuum distillation)

Fig. 9 Unusual, yet proper uses of the adapter with lots of names.

Fig. 10   The glassless glass adapter.

THINGS **NOT** TO DO WITH THE ADAPTER WITH LOTS OF NAMES

Forgetting the Glass

Look, the Corning people went to a lot of trouble to turn out a piece of glass (Fig. 10) that fits perfectly in *both* a glass joint *and* a rubber adapter, so *use it!*

Inserting Adapter Upside Down

This one (Fig. 11) is really ingenious. If you're tempted in this direction, go sit in the corner and repeat over and over,

"Only glass joints fit glass joints."

6

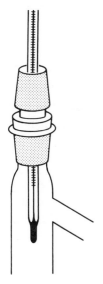

Fig. 11   The adapter stands on its head.

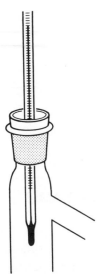

Fig. 12 The adapter on its head, without the head.

Inserting Adapter Upside down *sans* Glass

I don't know whether to relate this problem (Fig. 12) to glass forgetting or upside-downness, since it is both. Help me out. If I don't see you trying to use an adapter upside down without the glass, I won't have to make such a decision. So don't do it.

GREASING THE JOINTS

In all my time as an instructor, I've never had my students go overboard on greasing the joints, and they never got them stuck. Just lucky, I guess. Some instructors, however, use grease with a passion, and raise the roof over it. The entire concept of greasing joints is not as slippery as it may seem.

To Grease or Not to Grease

Generally you'll grease joints on two occasions. One, when doing vacuum work to make a tight seal that can be undone; the other, doing reactions with strong base that can etch the joints. Normally you don't have to protect the joints during acid or neutral reactions.

Preparation of the Joints

Chances are you've inherited a set of jointware coated with 47 semesters of grease. First wipe off any grease with a towel. Then soak a rag in any hydrocarbon solvent (hexane, ligroin, petroleum ether—and *no flames,* these burn like gasoline) and wipe the joint again. Wash off any remaining grease with a strong soap solution. You may have to repeat the hydrocarbon—soap treatments to get a clean, grease-free joint.

Into the Grease Pit (Fig. 13)

First, use only enough to do the job! Spread it thinly along the upper part of the joints only. Push the joints together with a twisting motion. The joint should turn clear from one third to one half the way down the joint. *At no time should the entire joint clear!* If it does, you have *too much grease* and must start back at *Preparation of the Joints.*

Don't interrupt the clear band around the joint. This is called **uneven greasing,** and it will cause you headaches later on.

STORING STUFF AND STICKING STOPPERS

At the end of a grueling lab session, you're naturally anxious to leave. The reaction mixture is sitting in the joint flask, all through reacting for the day, waiting in anticipation for the next lab. You put the correct glass stopper in the flask, clean up, and leave.

The next time, the stopper is stuck!

Stuck but good! And you can probably kiss your flask, stopper, product, and grade goodbye!

Frozen!

Material has gotten into the glass joint seal, dried out, and cemented the flask shut. There are few good cures, but several excellent preventive medicines.

Corks!

Yes, corks. Old-fashioned, non-stick-in-the-joint corks.

If the material you have to store *does not attack cork,* this is the cheapest, cleanest method of closing off a flask.

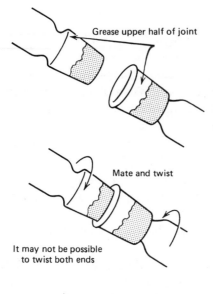

Grease upper half of joint

Mate and twist

It may not be possible
to twist both ends

Clear, unbroken band
of grease

Fig. 13 Greasing ground glass joints.

A well-greased glass stopper *can* be used for materials that attack cork, but *only* if the stopper has a good coating of stopcock grease. Unfortunately, this grease can get into your product.

Do not use rubber stoppers!

Organic liquids can make rubber stoppers swell up like beach balls. The rubber dissolves and ruins your product, and the stopper won't come out either. Ever.

The point is

Dismantle all ground glass joints
before you leave!

Other
Interesting
Equipment

7

An early edition of this book illustrated some equipment specific to the State University of New York at Buffalo, since that's where I was when I wrote it. It's now a few years later, and I realize that you can't make a comprehensive list.

Buffalo has an unusual "pear-shaped distilling flask" that I've not seen elsewhere. The University of Connecticut equipment list contains a "Bobbitt filter clip," which not many other schools have picked up.

So if you are disappointed that I don't have a list and drawing of every single piece of equipment in your drawer, I apologize. Only the most common organic lab equipment is covered here. Ask your instructor "Whattizzit?" if you do not know.

I assume that you remember Erlenmeyer flasks and beakers and such from the freshman lab. I'll discuss the other apparatus as it comes up in the various techniques. This might force you to read this book before you start lab.

The
Other
Equipment

8

Check out Figs. 14 and 15. Not all the mysterious doodads in your laboratory drawer are shown, but the more important are.

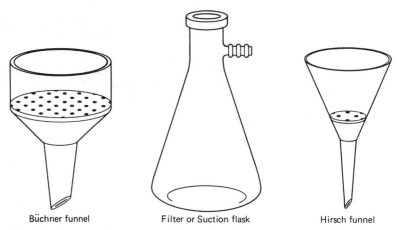

Büchner funnel   Filter or Suction flask   Hirsch funnel

Fig. 14 Some stuff from your lab drawer.

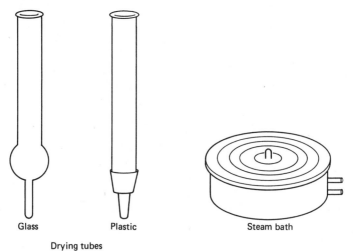

Glass   Plastic   Steam bath

Drying tubes

Fig. 15 More stuff from your lab drawer.

The Melting Point Experiment

9

A **melting point** is the temperature at which the first crystal just starts to melt until the temperature at which the last crystal just disappears. Thus the melting point (abbreviated M.P.) is actually a **melting range.** You should report it as such, even though it is *called* a melting point, for example, M.P. 147–149°C.

People always read the phrase as melting *point* and never as *melting* point. There is this uncontrollable, driving urge to report one number. No matter how much I've screamed and shouted at people *not* to report one number, they almost always do. It's probably because handbooks list only one number, the upper limit.

Generally, melting points are taken for two reasons.

1. ***Determination of purity.*** If you take a melting point of your compound and it starts melting at 60°C and doesn't finish until 180°C, you might suspect that something is wrong. A melting range *greater than 2°C* usually indicates an impure compound. (As with all rules, there are exceptions. There aren't many to this one, though.)

2. ***Identification of unknowns.***
   a. If you have an unknown solid, take a melting point. Many books (ask your instructor) contain tables of melting points and lists of compounds that may have a particular melting point. One of them may be your unknown. You may have 123 compounds to choose from. A little difficult, but that's not all the compounds in the world. Who knows? Give it a try. If nothing else, you know the melting point.
   b. Take your unknown and mix it *thoroughly* with some chemical you think might be your unknown. You might not get a sample of it, but you can ask. Shows you know something. Then:
      (i) If the mixture melts at a *lower* temperature, over a *broad range,* your unknown is NOT the same compound.
      (ii) If the mixture melts at the *same temperature, same range,* it's a good bet it's the *same compound.* Try another one, though, with a different ratio of your unknown and this compound just to be sure. A *lower* melting point with a *sharp range* would be a special point called a **eutectic mixture,** and you, with all the other troubles in lab, just might accidentally hit it. On lab quizzes, this is called

"Taking a mixed melting point."

Actually, "taking a mixture melting point," the melting point of a mixture, is more grammatically correct. But I have seen this expressed both ways.

SAMPLE PREPARATION

You usually take melting points in thin, closed-end tubes called **capillary tubes.** They are also called **melting point tubes** or even **melting point capillaries.** The terms are interchangeable, and I'll use all three.

Sometimes you may get a supply of tubes that are open on *both ends!* You don't just use these as is. Light up a burner, and close off one end, *before* you start. Otherwise your sample will fall out of the tube (see "Closing off Melting Point Tubes," below).

Take melting points on *dry, solid* substances ONLY, *never* on liquids or solutions of solids *in* liquids or on wet or even damp solids.

Only on dry solids!

To help dry damp solids, place the damp solid on a piece of filter paper and fold the paper around the solid. Press. Repeat until the paper doesn't get wet. Yes, you may have to use fresh pieces of paper. Try not to get filter paper fibers in the sample, OK?

Occasionally, you may be tempted to dry solid samples in an oven. *Don't*—unless you are specifically instructed to. I know some students who have decomposed their products in ovens and under heat lamps. With the time they save quickly decomposing their product, they can repeat the entire experiment.

Loading the Melting Point Tube

Place a small amount of *dry* solid on a new filter paper (Fig. 16). Thrust the open end of the capillary tube into the middle of the pile of material. Some solid should be trapped in the tube. Turn the tube over, closed end down. Remove any solid sticking to the outside. The solid must now be packed down.

Traditionally, the capillary tube, turned upright with the *open end up,* is stroked with a file or tapped on the benchtop. Unless done *carefully,* these operations *may break the tube.* A safer method is to drop the tube, *closed*

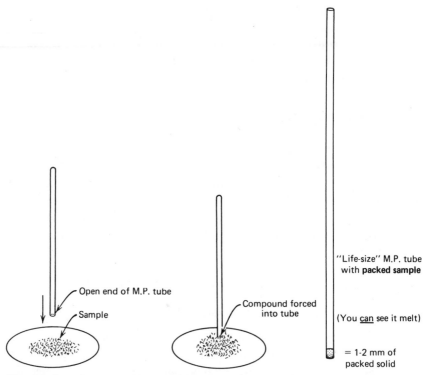

"Life-size" M.P. tube
with **packed sample**

(You <u>can</u> see it melt)

Open end of M.P. tube

Sample

Compound forced
into tube

= 1-2 mm of
packed solid

Fig. 16 Loading a melting point tube.

end down, through a length of glass tubing. You can even use your condenser or distilling column for this purpose. When the capillary strikes the benchtop, the compound will be forced into the closed end. You may have to do this several times. If there is not enough material in the M.P. tube, thrust the open end of the tube into the mound of material and pack it down again. Use your own judgment; consult your instructor.

**Use the smallest amount of material
that can be seen to melt.**

Closing off Melting Point Tubes

If you have melting point tubes that are *open at both ends,* and you try to take a melting point with one, it should come as no surprise when your compound falls out of the tube. You'll have to *close off one end* to keep your

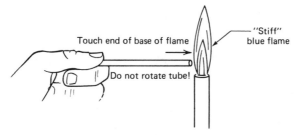

Touch end of base of flame →

"Stiff" blue flame

Do not rotate tube!

Fig. 17 Closing off an M.P. tube with a flame.

sample from falling out (Fig. 17). So fire up a burner and get a "stiff" small blue flame. SLOWLY touch the end of the tube to the side of the flame, and hold it there. You should get a yellow sodium flame, and the tube will close up. There is no need to rotate the tube. Remember, *touch—just touch*—the edge of the flame, and hold the tube there. Don't feel you have to push the tube way into the flame.

MELTING POINT HINTS

1. Use only the smallest amount that you can see melt. Larger samples will heat unevenly.
2. Pack down the material as much as you can. Left loose, the stuff will heat unevenly.
3. Never remelt any sample. They may undergo nasty chemical changes such as oxidation, rearrangement, and decomposition.
4. Make up more than one sample. One is easy, two is easier. If something goes wrong with one, you have another. Duplicate, even triplicate, runs are common.

THE MEL-TEMP APPARATUS

9

The Mel-Temp apparatus (Fig. 18) substitutes for the Thiele tube or open beaker and hot oil methods (see "Using the Thiele Tube," p. 40). Before you use the apparatus, there are a few things you should look for.

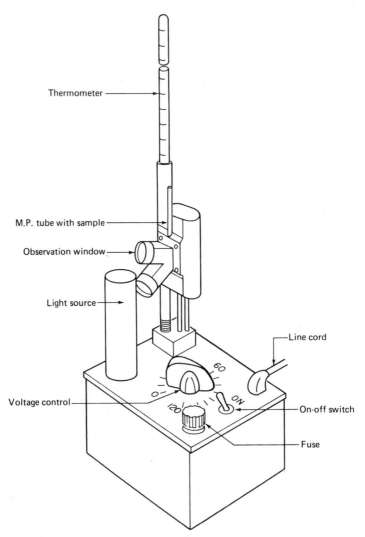

Thermometer

M.P. tube with sample

Observation window

Light source

Line cord

Voltage control

On-off switch

Fuse

Fig. 18 The Mel-Temp apparatus.

1. ***Line cord.*** Brings a.c. power to unit. Should be plugged into a live wall socket [see J.E. Leonard and L.E. Mohrmann, *J. Chem. Educ.*, **57,** 119 (1980), for a modification in the wiring of older units, to make them less lethal. It seems that even with the three-prong plug, there can still be a shock hazard. *Make sure your instructor knows about this!*].

2. ***On–off switch.*** Turns the unit on or off.
3. ***Fuse.*** Provides electrical protection for the unit.
4. ***Voltage control.*** Controls the *rate* of heating, *not the temperature!* The higher the setting, the faster the temperature rise.
5. ***Light source.*** Provides illumination for samples.
6. ***Eyepiece.*** Magnifies the sample (Fig. 19).
7. ***Thermometer.*** Gives temperature of sample, and upsets the digestion when you're not careful and you snap it off in the holder.

OPERATION OF THE MEL-TEMP APPARATUS

1. *Imagine yourself getting burned if you're not careful.* Never assume that the unit is cold.
2. Place loaded M.P. tube in one of the three channels in the opening at the top of the unit (Fig. 19).
3. Set the voltage control to zero if necessary. There are discourteous folk who do not reset the control when they finish using the equipment.

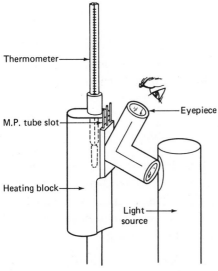

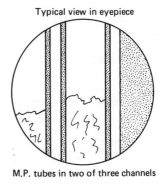

Typical view in eyepiece

M.P. tubes in two of three channels

Fig. 19 Closeup of the viewing system.

4. Turn on–off switch to **ON.** The light source should illuminate the sample. If not, call for help.
5. Now science turns into art. Set the **voltage control** to *any* convenient setting. The point is to get up to *within 20°C* of the *supposed* melting point. Yep, that's right. If you have no idea what the melting point is, it may require several runs as you keep skipping past the point with a temperature rise of 5–10°C per minute. A convenient setting is *40.* This is just a suggestion, not an article of faith.
6. After you've melted a sample, *throw it away!*
7. Once you have an idea of the melting point (or looked it up in a handbook or were told), *get a fresh sample,* and bring the temperature up quickly at about *5–10°C per minute* to *within 20°C* of this approximate melting point. Then turn down the *voltage* control to get a *2°C per minute rise.* Patience!
8. When the first crystals *just start to melt,* record the temperature. When the *last crystal just disappears,* record the temperature. If both points appear to be the same, either the sample is extremely pure or the temperature rise was *too fast.*
9. Turn the on–off switch to *OFF.* You can set the voltage control to zero for the next person.
10. Remove all capillary tubes.

Never use a wet rag or sponge to quickly cool off the heating block. This might permanently warp the block. You can use a cold metal block to cool it if you're in a hurry. Careful. If you slip, you may burn yourself.

THE FISHER–JOHNS APPARATUS

The Fisher–Johns apparatus (Fig. 20) is different in that you don't use capillary tubes to hold the sample. Instead, you sandwich your sample between two round microscope cover slides (thin windows of glass) on a heating block. This type of melting point apparatus is called a **hot stage.** It comes complete with spotlight. Look for the following.

1. *Line cord* (at the back). Brings a.c. power to unit. Should be plugged into a live wall socket.

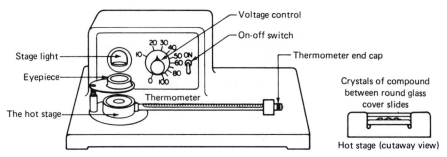

Fig. 20 The Fisher–Johns apparatus.

2. **On–off switch.** Turns the unit on or off.
3. **Fuse** (also at the back). Provides electrical protection for the unit.
4. **Voltage control.** Controls the *rate* of heating, *not the temperature!*
    The higher the setting, the faster the temperature rise.
5. **Stage light.** Provides illumination for samples.
6. **Eyepiece.** Magnifies the sample.
7. **Thermometer.** Gives temperature of sample.
8. **Thermometer end cap.** Keeps thermometer from falling out. If the
    cap becomes loose, the thermometer tends to go belly-up, and the
    markings turn over. Don't try to fix this while the unit is hot. Let it cool
    so you won't get burned.
9. **Hot stage.** This is the heating block that samples are melted on.

OPERATION OF THE FISHER–JOHNS APPARATUS

1. *Don't assume that the unit is cold.* That is a good way to get burned.
2. Keep your grubby fingers off the cover slides. Use tweezers or forceps.
3. Place a clean round glass cover slide in the well on the hot stage.
    Never melt any samples directly on the metal stage. Ever!
4. Put a few crystals on the glass. Not too many. As long as you can see
    them melt, you're all right.
5. Put another cover slide on top of the crystals to make a sandwich.
6. Set the voltage control to zero if it's not already there.

9

7. Turn on–off switch to **ON.** The light source should illuminate the sample. If not, call for help!
8. Now science turns into art. Set the *voltage control to any* convenient setting. The point is to get up to *within 20°C* of the *supposed* melting point. Yep, that's right. If you have no idea what the melting point is, it may require several runs as you keep skipping past the point with a temperature rise of 5–10°C per minute. A convenient setting is *40.* This is just a suggestion, not an article of faith.
9. After you've melted a sample, let it cool, and remove the sandwich of sample and cover slides. *Throw it away!* Use an appropriate waste container.
10. Once you have an idea of the melting point (or looked it up in a handbook or were told), *get a fresh sample,* and bring the temperature up quickly at about *5–10°C per minute* to within *20°C* of this approximate melting point. Then turn down the *voltage* control to get a *2°C per minute rise.* Patience!
11. When the first crystals *just start to melt,* record the temperature. When the last crystal *just disappears,* record the temperature. If both points appear to be the same, either the sample is extremely pure or the temperature rise was *too fast.*
12. Turn the on–off switch to *OFF.* Now set the voltage control to zero.
13. Let the stage cool, then remove the last sandwich.

There are a few more pieces of electric melting point apparatus around, and many of them work the same. A **sample holder, magnifying eyepiece,** and **voltage control** are common, and an apparently essential feature of these devices is that dial markings are almost *never* temperature settings. That is, a setting of *60* will not give a temperature of 60°C but probably much higher.

USING THE THIELE TUBE

With the Thiele tube (Fig. 21) you use hot oil to transfer heat evenly to your sample in a melting point capillary, just like the metal block of the Mel-Temp apparatus does. You heat the oil in the sidearm and it expands. This

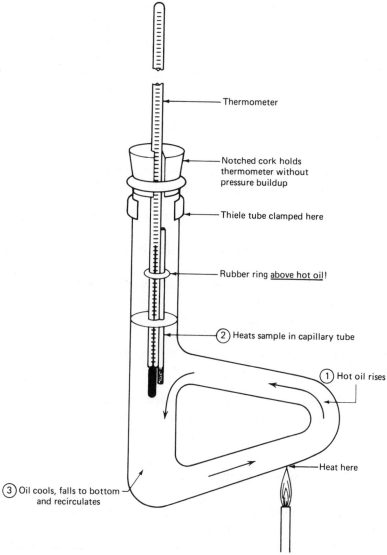

Thermometer

Notched cork holds thermometer without pressure buildup

Thiele tube clamped here

Rubber ring above hot oil!

(2) Heats sample in capillary tube

(1) Hot oil rises

Heat here

(3) Oil cools, falls to bottom and recirculates

Fig. 21 Taking melting points with the Thiele tube.

9

hot oil goes up the sidearm, warming your sample and thermometer as it touches them. Now, the oil is cooler and it falls to the bottom of the tube where it is heated again by a burner. This cycle goes on automatically as you do the melting point experiment in the Thiele tube.

Don't get any water in the tube or when you heat the tube the water may boil and throw hot oil out at you. Let's start from the beginning.

Cleaning the Tube

This is a bit tricky, so don't do it unless your instructor says so. Also, check with him *before* you put fresh oil in the tube.

1. Pour the old oil out into an appropriate container and let the tube drain.
2. Use a hydrocarbon solvent (hexane, ligroin, petroleum ether—and *no flames!*) to dissolve the oil that's left.
3. Get out the old soap and water and elbow grease, clean the tube, and rinse it out really well.
4. Dry the tube in a drying oven (usually >100°C) thoroughly. Carefully take it out of the oven and let it cool.
5. Let your instructor examine the tube. If he says it's OK, *then* add some fresh oil.

Watch it. First, *no water*. Second, don't overfill the tube. Normally, the oil expands as you heat the tube. If you've overfilled the tube, oil will crawl out and get you.

Getting the Sample Ready

Here you use a loaded melting point capillary tube (see "Loading the Melting Point Tube" above) and attach it directly to the thermometer. The thermometer, unfortunately, has bulges; there are some problems, and you may snap the tube while attaching it to the thermometer.

1. Get a thin rubber ring, or cut one from a piece of rubber tubing.
2. Put the *bottom* of the loaded M.P. tube *just above* the place where the

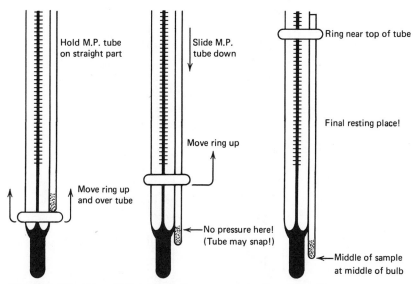

Fig. 22 Attaching M.P. tube to thermometer without disaster.

thermometer constricts (Fig. 22), and carefully roll the rubber ring onto the M.P. tube.

3. Reposition the tube so that the sample is near the center of the bulb, and the rubber ring is near the open end. *Make sure the tube is vertical.*

Dunking the Melting Point Tube

There are more ways of keeping the thermometer suspended in the oil than I care to list. You can cut or file a notch on the side of the cork, drill a hole, and insert the thermometer. (*Careful!*) Finally, cap the Thiele tube (Fig. 21). The notch is there so that pressure will not build up as the tube is heated. *Keep the notch open, or the setup may explode.*

But this requires drilling or boring corks, something you try to avoid (why have ground glass jointware in the undergraduate lab?) You can *gently* hold a thermometer and a cork in a clamp (Fig. 23). Not too much pressure, though!

Finally, you might put the thermometer in the **thermometer adapter** and suspend that, clamped gently by the rubber part of the adapter, not by the ground glass end (Fig. 23). Clamping ground glass will score the joint.

9

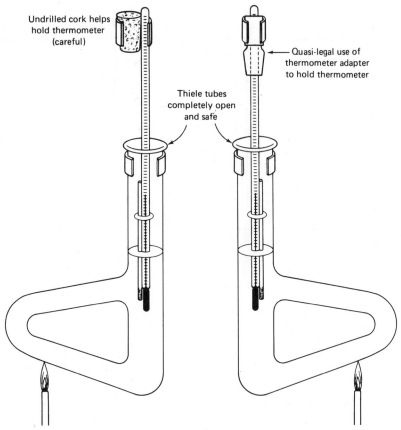

Fig. 23 Safely suspended thermometer with Thiele tube.

Heating the Sample

The appropriately clamped thermometer is set up in the Thiele tube as in Fig. 21. Look at this figure *now* and remember to heat the tube *carefully—always carefully*—at the elbow. Then:

1. If you don't know the melting point of the sample, heat the oil fairly quickly, *but no more than 10°C per minute* to get a rough melting point. And it will be rough indeed, since the temperature of the thermometer usually lags that of the sample.

2. After this sample has melted, lift the thermometer with attached sample tube *carefully* (*it may be HOT*) by the thermometer up at the clamp, until they are *just out of the oil*. This way the thermometer and sample can cool, and the hot oil can drain off. Wait for the thermometer to cool to about room temperature before you remove it entirely from the tube. Wipe off some of the oil, reload a melting point tube (*never remelt melted samples*), and try again. And heat at 2°C per minute this time.

9

Recrystallization

10

The essence of a recrystallization is a **purification.** Messy, dirty compounds are cleaned up, purified, and can then hold their heads up in public again. The sequence of events you use will depend a lot on how messy your crude product is and just how soluble it will be in various solvents.

In any case, you'll have to remember a few things.

1. Find a solvent that will *dissolve the solid while hot.*
2. The same solvent *should not dissolve it while cold.*
3. The *cold solvent* must keep impurities dissolved in it *forever or longer.*

This is the major problem. And it requires some experimentation. That's right! Once again, art over science. Usually, you'll know what you should have prepared, so the task is easier. It requires a trip to **your notebook,** and possibly, a **handbook** (see Chapter 4, "Keeping a Notebook," and Chapter 5, "Interpreting a Handbook"). You have the data on the solubility of the compound in your notebook. What's that you say? *You don't have the data in your notebook?* Congratulations, you get the highest F in the course.

Information in the notebook (which came from a handbook) for your compound might be, for alcohol (meaning *ethyl* alcohol), **s.h.** Since this means **s**oluble in **h**ot alcohol, it implies insoluble in cold alcohol (and you wondered what the **i** meant). Then alcohol is probably a good solvent for recrystallization of that compound. Also, check on the **color** or **crystalline form.** This is important since

1. A color in a supposedly white product is an impurity.
2. A color in a colored product is *not* an impurity.
3. The *wrong color* in a product is an impurity.

You can usually assume that impurities are present in small amounts. Then you don't have to guess what possible impurities might be present or what they might be soluble or insoluble in. If your sample is really dirty, the assumption can be fatal. This doesn't usually happen in an undergraduate lab, but you should be aware of the possibility.

If the solubility data for your compound are not in handbooks, place 0.1 g of the solid in a test tube with 3 ml of solvent. The compound is soluble only if *all* the compound dissolves. By definition, if any crystals don't dissolve, the compound is insoluble. Yes, you may heat the solvent (*no flames!*) if you want. Then you'll have the solubility in hot solvent.

Here are some general rules to follow for purifying any solid compound.

1. Put the solid in an Erlenmeyer flask, not a beaker. If you recrystallize compounds in beakers, you may find the solid climbing the walls of the beaker to get at you as a reminder. A 125-ml Erlenmeyer usually works. Your solid should look comfortable in it, neither cramped nor with too much space. You probably shouldn't fill the flask more than one fifth to one fourth full.

2. Heat a large quantity of a proven solvent (see above) to the boiling point, and *slowly add the hot solvent. Slowly!* A word about solvents: *Fire! Solvents burn! No flames!* A hot plate would be better. You can even heat solvents on a *steam or water bath.* But—*no flames!*

3. Carefully add the hot solvent to the solid to just dissolve it. This can be tricky, since hot solvents evaporate, cool down, and so on. Ask your instructor.

4. Add a slight excess of the hot solvent (5–10 ml) to keep the solid dissolved.

5. If the solution is only slightly colored, the impurities will stay in solution. Otherwise, the big gun, **activated charcoal,** may be needed (see p. 55). Remember, if you were working with a colored compound, it would be silly to try to get rid of all the color, since you would get rid of all the compound and probably all your grade.

6. Keep the solvent hot (*not boiling*) and look carefully to see if there is any trash in the sample. This could be old boiling stones, sand, floor sweepings, and so on. Nothing you'd want to bring home to meet the folks. *Don't confuse real trash with undissolved good product!* If you add more hot solvent, good product will dissolve, and trash will not. If you have trash in the sample, do a **gravity filtration** (see below).

7. Let the Erlenmeyer flask and the hot solution cool. Slow cooling gives better crystals. Garbage doesn't get trapped in them. But this can take what seems to be an interminable length of time. (I know, the entire lab seems to take an interminable length of time.) So, after the flask cools and it's just *warm* to the touch, then put the flask in an ice-water

10

bath to cool. *Watch it!* Flasks have a habit of turning over in water baths and letting water destroy all your hard work! Also, a really hot flask will shatter if plunged into the ice bath, so again, watch it.

8. When you're through cooling, filter the crystals on a **Büchner funnel** (see p. 53).
9. Dry them, and take a melting point, as described in Chapter 9.

GRAVITY FILTRATION

If you find yourself with a flask full of hot solvent, your product dissolved in it, along with all sorts of trash, this is for you. You'll need more hot solvent, a ringstand with a ring attached, possibly a clay triangle, some filter paper, a clean, dry flask, and a **stemless funnel.** Here's how **gravity filtration** works.

1. Fold up a **filter cone** out of a piece of filter paper (Fig. 24). It should fit nicely, within a single centimeter or so of the top of the funnel. For those who wish to filter with more panache, try using **fluted filter paper** (see "world famous fan-folded fluted filter paper," Fig. 34, below).
2. Get yourself a **stemless funnel,** or, at least, a **short-stemmed funnel.** Why? Go ahead and *use* a stem funnel and watch the crystals come out in the stem as the solution cools, blocking up the funnel (Fig. 25).
3. Put the **filter paper cone** in the **stemless funnel.**
4. Support this in a ring attached to a ringstand (Fig. 26). If the funnel is

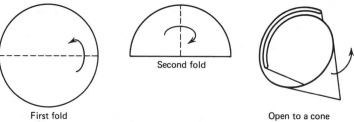

First fold   Open to a cone

Second fold

Fig. 24 Folding filter paper for gravity filtration.

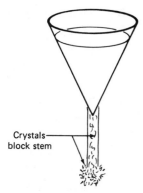

Crystals
block stem

Fig. 25 Too long a funnel stem—Oops!

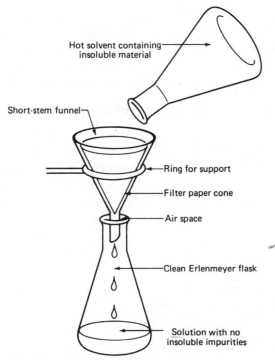

Hot solvent containing
insoluble material

Short-stem funnel

Ring for support

Filter paper cone

Air space

Clean Erlenmeyer flask

Solution with no
insoluble impurities

Fig. 26 The gravity filtration setup with a funnel that fits the iron ring.

10

too small, and you think it could fall through the ring, you may be able to get a **wire** or **clay triangle** to support the funnel in the ring (Fig. 27).

5. Put a new, *clean, dry flask* under the funnel to catch the hot solution as it comes through. All set?

6. Get that flask with the solvent, product, and trash hot again. (*No flames!*) You should get some fresh, clean solvent hot as well. (*No flames!*)

7. Carefully pour the hot solution into the funnel. As it is, some solvents evaporate so fast that product will probably come out on the filter paper. It is often hard to tell the product from the insoluble trash. Then—

8. Wash the filter paper down with *a little hot solvent*. The product will redissolve. The trash won't.

9. You now let the *trash-free* solution cool, and clean crystals should come out. Since you have probably added solvent to the solution, *don't be surprised if no crystals come out of solution. Don't panic either!* Just boil away some of the solvent, let your solution cool, and wait for the crystals again. If they *still* don't come back, repeat the boiling.

Do not boil to dryness!

Somehow, lots of folk think recrystallization means dissolving the solid, then boiling away all the solvent to dryness. *No!* There must be a way to convince these lost souls that *the impurities will deposit on the crystals.* After the solution has cooled, crystals come out, sit on the bottom of the flask, and *must be covered by solvent!* There must be enough solvent to keep those nasty impurities dissolved and off the crystals.

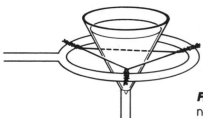

Fig. 27 A wire triangle holding a small funnel in a large iron ring for the gravity filtration setup.

THE BÜCHNER FUNNEL AND FILTER FLASK

The **Büchner funnel** (Fig. 14) is used primarily for separating crystals of product from the liquid above them. If you have been *boiling your recrystallization solvents dry, you should be horsewhipped* and forced to reread the sections on recrystallization!

1. Get a piece of filter paper large enough to cover all the holes in the bottom plate, yet *not* curl up the sides of the funnel. It is placed *flat* on the plate (Fig. 28).
2. A **rubber stopper** or **filter adapter** is used to stick the funnel into the top of the **filter flask.** This filter flask, often called a **suction flask,** is a very heavy-walled flask with a sidearm on the neck. A piece of heavy-walled tubing connects this flask to the **water trap** (see Fig. 30). Clamp the flask and funnel to a ringstand. The Büchner

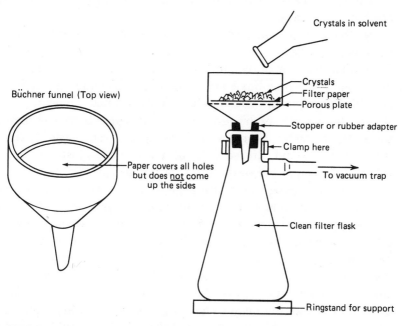

Crystals in solvent

Büchner funnel (Top view)

Crystals
Filter paper
Porous plate

Stopper or rubber adapter

Clamp here

Paper covers all holes but does <u>not</u> come up the sides

To vacuum trap

Clean filter flask

Ringstand for support

Fig. 28   The Büchner funnel at home and at work.

10

funnel makes the setup top-heavy and prone to be prone—and broken.

3. The **water trap** is in turn connected to a source of vacuum, most likely a **water aspirator** (Fig. 29 shows one).

4. The faucet on the **water aspirator** should be turned on *full blast!* This should suck down the filter paper, which you now *wet with some of the cold recrystallization solvent.* This will make the paper stick to the plate.

5. Swirl and pour the crystals and solvent *slowly, directly into the center of the filter paper,* as if to build a small mound of product there. *Slowly!* Don't flood the funnel by filling it right to the brim and waiting for the level to go down. If you do that, the paper may float up, ruining the whole setup.

6. Use a very small amount of the same cold recrystallization solvent and a spatula to remove any crystals left in the flask. Then you can use some of the *fresh, cold recrystallization solvent* and slowly pour it over the crystals to wash away any old recrystallization solvent and dissolved impurities.

7. Leave the aspirator on, and let air pass through the crystals to help them dry. You can put a thin rubber sheet, a **rubber dam,** over the funnel. The vacuum pulls it in, and the crystals are pressed clean and dry. And you won't have air or moisture blowing through, and possibly decomposing, your product. Rubber dams are neat.

8. When the crystals are dry, *and you have a* **water trap,** just turn off the water aspirator. Water won't back up into your flask. (If you've been foolhardy and filtered without a water trap, just remove the rubber tube connected to the filter flask sidearm; Fig. 28).

9. At this point, you may have a *cake of crystals* in your Büchner funnel. The easiest way to handle this is to *carefully lift the cake* of crystals out of the funnel *along with the filter paper,* plop the whole thing onto a larger piece of filter paper, and let it all dry overnight. If you are pressed for time, *scrape the damp filter cake from the filter paper, but don't scrape any filter paper fibers into the crystals.* Repeatedly press the crystals out between dry sheets of filter paper, changing sheets until the crystals no longer show any solvent spot after pressing. Those who use **heat lamps** may find their white crystalline product turning into instant charred remains.

10. When your cake is *completely dried,* weigh a vial, put in the product, and weigh the vial again. Subtracting the weight of the vial from the weight of the vial and sample will give the weight of the product. This **weighing by difference** is easier and less messy than weighing the crystals directly on the balance. This weight should be included in the label on your **product vial** (see Chapter 22, "On Products").

ACTIVATED CHARCOAL

Activated charcoal is ultrafinely divided carbon with lots of places to suck up big, huge, polar, colored impurity molecules. Unfortunately, if you use too much, it'll suck up *your product!* And, if your product was white or yellow, it'll have a funny gray color from the excess charcoal. Sometimes the impurities are untouched, and only the product gets absorbed. Again, it's a matter of trial and error. Try not to use too much. Suppose you've got a *hot solution* of some solid, *and the solution is highly colored.* Well,

1. First *make sure your product should not* **be** *colored!*
2. Take the flask with your filthy product off the heat and swirl the flask. This dissipates any superheated areas so that when you add the activated charcoal, the solution doesn't foam out of the flask and onto your shoes.
3. *Add the activated charcoal.* Put a small amount, about the size of a pea, on your spatula, then throw the charcoal in. Stir. The solution may turn black. Stir and heat.
4. Set up the **gravity filtration** and filter off the carbon. It is especially important to *wash off any product caught on the charcoal,* and it is really hard to see anything here. You should take advantage of **fluted filter paper.** It should give a more efficient filtration.
5. Yes, have some extra fresh solvent heated as well. You'll need to add a few milliliters of this to the hot solution to help keep the crystals from coming out on the filter paper. And you'll need more to help wash the crystals off the paper when they come out on it anyway.
6. This solution should be *much cleaner* than the original solution. If not, you'll have to *add charcoal and filter again.* There is a point of diminishing returns, however, and one or two treatments is usually all you should do. Get some guidance from your instructor.

10

Your solid products should not be gray. Liquid products (yes, you can do liquids!) will let you know that you didn't get all the charcoal out. Often, you can't see charcoal contamination in liquids while you're working with them. The particles stay suspended for awhile, but after a few days, you can see a layer of charcoal on the bottom of the container. Sneaky, those liquids. By the time the instructor gets to grade all the products—voilà—the charcoal has appeared!

THE WATER ASPIRATOR: A VACUUM SOURCE

Sometimes you'll need a vacuum for special work like **vacuum distillation** and **vacuum filtration** as with the Büchner funnel (above). An inexpensive source of vacuum is the **water aspirator** (Fig. 29).

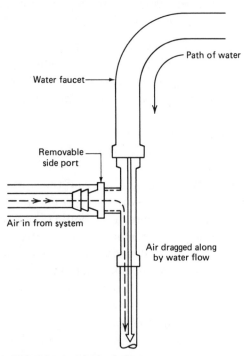

Fig. 29 A water aspirator.

When you turn the water on, the water flow draws air in from the side port on the aspirator. The faster the water goes through, the faster the air is drawn in. Pretty neat, huh? I've shown a plastic aspirator, but many of the older metal varieties are still around.

You may have to pretest some aspirators before you find one that will work well. It'll depend on the water pressure in the pipes, too. Even the number of people using aspirators on the same water line can affect the performance of these devices. You can test them by going to an aspirator and turning the faucet on *full blast*. It does help to have a sink under the aspirator. If water leaks out the side port, *tell your instructor and find another aspirator*. Wet your finger and place it over the hole in the side port to feel if there is any vacuum. If there is *no vacuum,* tell your instructor and find another aspirator. Some of these old, wheezing aspirators have a very weak vacuum. You must decide for yourself whether the suction is "strong enough." There should be a **splash guard** or rubber tubing leading the water stream directly into the sink to keep water from going all over the room. If you check and don't find such protection, see your instructor. All you have to do with a fully tested and satisfactory aspirator is hook it up to the **water trap.**

THE WATER TRAP

Every year I run a chem lab and when someone is doing a **vacuum filtration,** suddenly I'll hear a scream and a moan of anguish as water backs up into someone's filtration system. Usually there's not much damage, since the filtrate in the suction flask is generally thrown out. For **vacuum distillations,** however, this **suck-back** is disaster. It happens whenever there's a pressure drop on the water line big enough to cause the flow to decrease, so that there is a *greater vacuum in the system than in the aspirator.* Water, being water, flows into the system. Disaster.

So, for your own protection, make up a **water trap** from some stoppers, rubber tubing, a thick-walled Erlenmeyer or filter flask, and a screw clamp (Fig. 30). Two versions are shown. I think the setup using the filter flask is more flexible. The screw clamp allows you to let air into your setup at a

10

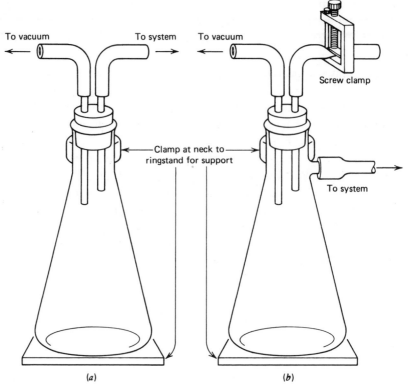

Fig. 30 A couple of water traps hanging around.

controlled rate. You might clamp the water trap to a ringstand when you use it. The connecting hoses have been known to flip unsecured flasks two out of three times.

WORKING WITH A MIXED-SOLVENT SYSTEM—THE GOOD PART

If, after sufficient agony, you cannot find a single solvent to recrystallize your product from, you may just give up and try a *mixed-solvent system*. Yes, it does mean you mix more than one solvent and *recrystallize using the mixture*. It should only be so easy. Sometimes you are told what the mixture is and the correct proportions. Then it is easy.

For an example, I could use "solvent 1" and "solvent 2," but that's clumsy. So I'll use the ethanol–water system and point out the interesting stuff as I go along.

The Ethanol–Water System

If you look up the solubility of water in ethanol (or ethanol in water) you find an ∞. This means they mix in all proportions. Any amount of one dissolves completely in the other, no matter what. Any volumes, any weights. You name it. The special word for this property is **miscibility.** Miscible solvent systems are the kinds that you should use as mixed solvents. They keep you out of trouble. You'll be adding amounts of water to the ethanol and ethanol to the water. If the two weren't miscible, they might begin to separate and form two layers as you changed the proportions. Then you'd have REAL trouble. So, go ahead. You *can* work with mixtures of solvents that aren't miscible. But don't say you haven't been warned.

The ethanol–water mixture is so useful because

1. *At high temperatures, it behaves like alcohol!*
2. *At low temperatures, it behaves like water!*

From this, you should get the idea that it would be good to use a mixed solvent to recrystallize compounds that are *soluble in alcohol* yet *insoluble in water.* You see, each solvent, alone, cannot be used. If the material is soluble in the alcohol, not many crystals come back from alcohol alone. If the material is insoluble in water, you cannot even begin to dissolve it. So, you have a *mixed solvent* with the best properties of *both* solvents. To perform a *mixed-solvent recrystallization* you

1. Dissolve the compound in the smallest amount of *hot ethanol*.
2. Add *hot water* until the solution turns cloudy. This **cloudiness** is *tiny crystals of compound coming out of solution.* Heat this solution to dissolve the crystals. If they do not dissolve completely, add *a very little hot ethanol* to force them back into solution.
3. Cool and collect the crystals on a **Büchner funnel.**

10

Any solvent pair that behaves the same way can be used. The addition of one hot solvent to another can be tricky. It is *extremely dangerous* if the boiling points of the solvents are very different. For the *water–methanol mixed solvent,* if 95°C water hits *hot methanol* (B.P. 65.0°C), watch out!

A MIXED-SOLVENT SYSTEM—THE BAD PART

Every silver lining has a cloud. More often than not, compounds "recrystallized" from a mixed-solvent system don't form crystals. Your compound may form an *oil* instead.

Oiling out is what it's called; more work is what it means. Compounds usually oil out if *the boiling point of the recrystallization solvent is higher than the melting point of the compound,* though that's not the only time. In any case, if the oil solidifies, the impurities are trapped in the now solid "oil," and you have to purify the solid again.

Don't think you won't ever get oiling out if you stick to single, unmixed solvents. It's just that with two solvents, you have a greater chance of hitting upon a composition that will cause this.

Temporarily, you can

1. Add more solvent. If it's a mixed-solvent system, try adding more of the solvent the solid is NOT soluble in. Or add more of the OTHER solvent. No contradiction. The point is to *change the composition.* Single solvent or mixed solvent, changing the composition is one way out of this mess.
2. Redissolve the oil by heating, then shake up the solution as it cools and begins to oil out. When these smaller droplets finally freeze out, they may form crystals that are relatively pure. They may not. You'll probably have to clean them up again. Just don't use the same recrystallization solvent.

Sometimes, once a solid oils out, it doesn't want to solidify at all, and you might not have all day. Try removing a sample of the oil with an eyedropper or disposable pipet. Then get a glass surface (watch glass) and add a few drops of a solvent that the compound is known to be *insoluble* in (usually water). Then use the rounded end of a glass rod to *triturate the oil with the solvent*. **Trituration** can be described loosely as the beating of an oil into a crystalline solid. Then you can put these crystals back into the rest of the oil. Possibly they'll act as seed crystals and get the rest of the oil to solidify. Again, you'll still have to clean up your compound.

SALTING OUT

Sometimes you'll have to recrystallize your organic compound from water. No big deal. But sometimes your organic compound is more than ever so slightly insoluble in water, and not all the compound will come back. Solution? Salt solution! A pinch of salt in the water raises the **ionic strength.** There are now charged ions in the water. Some of the water that solvated your compound goes to be with the salt ions. Your organic compound does not particularly like charged ions anyway, so more of your organic compound comes out of solution.

You can dissolve about 36 g of common salt in 100 ml of cold water. That's the upper limit for salt. You can estimate how much salt you'll need to practically saturate the water with salt. Be careful though—if you use too much salt, you may find yourself collecting salt crystals along with your product (see also the application of salting out when you have to do an extraction; "Extraction Hints," p. 80).

WORLD FAMOUS FAN-FOLDED FLUTED FILTER PAPER

Some training in origami is *de rigeur* for chemists. It seems that the regular filter paper fold is inefficient, since very little of the paper is exposed. The idea is to **flute** or **corrugate** the paper, increasing the surface area in contact with your filtrate. You'll have to do this several times to get good at it.

Right here let's review the difference between **fold** and **crease.** Folding is folding; creasing is folding, then stomping on it, running fingers and fingernails over a fold over and over and over. Creasing so weakens the

10

paper, especially near the point, that it may break at an inappropriate time in the filtration.

1. Fold the paper in half, then in half again, then in half again (Fig. 31). Press on this wedge of paper to get the fold lines to stay, but *don't crease. Do this in one direction only.* Either always fold toward you or away from you, but not both.
2. Unfold this cone *twice* so it looks like a semicircle (Fig. 32), and put it down on a flat surface. Look at it and think for not less than two full minutes the first time you do this.
3. OK. Now try a "fan fold." You alternately fold, first in one direction then the other, every individual section of the semicircle (Fig. 33).
4. Open the fan and play with it until you get a fairly fluted filter cone (Fig. 34).

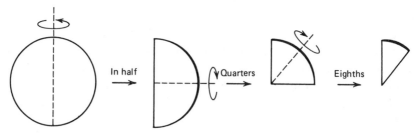

Fig. 31 Folding filter paper into eighths.

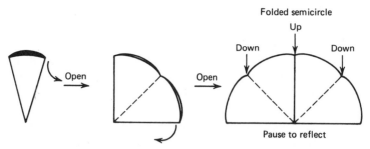

Fig. 32 Unfolding to a sort of bent semicircle.

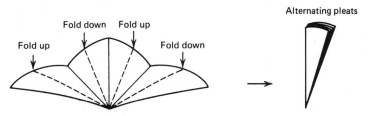

Fig. 33   Refolding to a fan.

5. It'll be a bit difficult, but try to find the two opposing sections that are NOT folded correctly. Fold them inward (Fig. 34), and you'll have a fantastic fan-folded fluted filter paper of your very own.

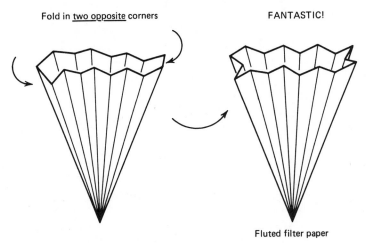

Fig. 34   Finishing the fluted fan.

P.S. For those with more money than patience, prefolded fan-folded fluted filter paper is available from suppliers.

10

Extraction

11

Extraction is one of the more complex operations you'll do in the organic chemistry lab. For this reason, I'll go over it especially slowly and carefully. Another term you'll see used simultaneously with **extraction** is **washing.** That's because extraction and washing are really the same operation, but each leads to a different end. How else to put this?

Let's make some soup. Put the vegetables, fresh from the store, in a pot. You run cold water in and over them to clean them and throw this water down the drain. Later, you run water in and over them to cook them. You keep this water—it's the soup.

Both operations are similar. Veggies in a pot in contact with water the first time is a **wash.** You remove unwanted dirt. *You washed with water.* The second time, veggies in a pot in contact with water is an **extraction.** You've **extracted** essences of the vegetables *into water.* Very similar operations; very different ends.

To put it a little differently,

You would extract good material from an impure matrix.

You would wash impurities from good material.

The vegetable soup preparation is a **solid-liquid extraction.** So is coffee making. You extract some component(s) of a solid directly into the solvent. You might do a solid–liquid extraction in lab as a separate experiment; *liquid–liquid extractions are routine.* They are so common that if you are told to do an extraction or a washing, *it is assumed* you will use *two liquids—two INSOLUBLE liquids—and a separatory funnel.* The separatory funnel, called a **sep funnel** by those in the know, is a special funnel that you can separate liquids in. You might look at the section on separatory funnels (p. 69) right now, then come back later.

Two insoluble liquids in a separatory funnel will form **layers;** one liquid will float on top of the other. You usually have compounds dissolved in these layers, and either the compound you want is *extracted from one to the other* or junk you don't want is *washed from one layer to the other.*

Making the soup, you have *no* difficulty deciding what to keep or what to throw away. First you throw the water away; later you keep it. But this can change. In a sep funnel, the layer you want to keep one time may not be the layer you want to keep the next time. Yet, if you throw one layer away prematurely, you are doomed.

NEVER-EVER LAND

11

**Never, never, never, never,
ever throw away any layer, until you are absolutely sure you'll
never need it again. Not very much of your product can be recovered from the sink trap!**

I'm using a word processor, so I can copy this warning over and over
again, but let's not get carried away. One more time, WAKE UP OUT
THERE!

**Never, never, never, never,
ever throw away any layer, until you are absolutely sure you'll
never need it again. Not very much of your product can be recovered from the sink trap!**

STARTING AN EXTRACTION

To do any extraction, you'll need two liquids, or solutions. *They must be
insoluble in each other.* **Insoluble** here has a practical definition:

**When mixed together,
the two liquids form two layers.**

One liquid will float on top of the other. A good example is ether and
water. Handbooks say that ether is slightly soluble in water. When ether
and water are mixed, yes, some of the ether dissolves; most of the ether just
floats on top of the water.

Really soluble or *miscible liquid pairs are no good* for extraction and
washing. When you mix then, *they will not form two layers!* In fact, they'll
mix in all proportions. A good example of this is acetone and water. What
kinds of problems can this cause? Well, for one, *you cannot perform any
extraction with two liquids that are miscible.*

Let's try it. A mixture of, say, some mineral acid (is HCl all right?) and
an organic liquid, "compound A," needs to have that acid washed out of it.
You dissolve the compound A–acid mixture in some acetone. It goes into
the sep funnel, and you now add water to wash out the acid.

Acetone is miscible in water.
No layers form! You lose!

Back to the lab bench. Empty the funnel. Clean it and dry it. Start over. This time, having called yourself several colorful names because you should have read this section thoroughly in the first place, you dissolve the compound A—acid mixture in ether and put it into the sep funnel. Add water, and *two layers form!* Now you can wash the acid from the organic layer to the water layer. The water layer can be thrown away.

Note that the acid went into the water, *then the water was thrown out!* So we call this a **wash.** If the water layer had been saved, we'd say the acid had been **extracted into the water layer.** It may not make sense, but that's how it is.

Review:

1. You *must have two insoluble liquid layers* to perform an extraction.
2. *Solids must be dissolved in a solvent,* and that solvent must be insoluble in the other extracting or washing liquid.
3. If you are washing or extracting an organic liquid, dissolve it into another liquid, *just like a solid,* before extracting or washing it.

So these terms, **extraction and washing,** are related. Here are a few examples.

1. Extract with ether. Throw ether together with the solution of product and pull out *only the product into the ether.*
2. Wash with 10% NaOH. Throw 10% NaOH together with the solution of product and pull out *everything but product into the NaOH.*
3. You can even extract with 10% NaOH.
4. You can even wash with ether.

So extraction is pulling out what you want
from all else!

Washing is pulling out all else
from what you want.

And please note—*you ALWAYS do the pulling from ONE LAYER INTO ANOTHER.* That's also *two immiscible liquids.*

You'll have to actually do a few of these things before you get the hang of it, but bear with me. When your head stops hurting, reread this section.

11

THE SEPARATORY FUNNEL

Before going on to some practical examples, you might want to know more about where all this washing and extracting is carried out. I've mentioned that it's a special funnel called a **separatory funnel** (Fig. 35) and that you can impress your friends by calling it a **sep funnnel.** Here are a few things you should know.

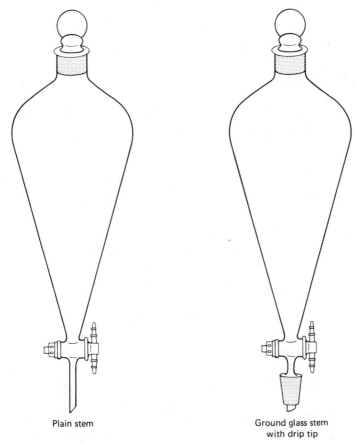

Plain stem

Ground glass stem
with drip tip

Fig. 35 Garden variety separatory funnels.

The Stopper

At the top of the sep funnel is a ⅀ glass stopper. There is a number, commonly ⅀22, possibly ⅀19/22, on the stopper head. *Make sure the head has a number and that the same number is marked on the funnel.* If this stopper is not so marked, you may find the product leaking over your shoes when you turn the sep funnel upside down. Try *not to grease this stopper* unless you plan to sauté your product. Unfortunately, these stoppers tend to get stuck in the funnel. The way out is to be sure you don't get the ground glass surfaces wet with product. How? Pour solutions into the sep funnel as carefully as you might empty a shotglass of Scotch into the soda. Maybe *use a funnel*. To confuse matters, I'll suggest you use a light coating of grease. Unfortunately, my idea of light and your idea of light may be different.

Consult your instructor!

The Glass Stopcock

This is the time-honored favorite of separatory funnel makers everywhere. There is a notch at the small end that contains either a rubber ring *or* a metal clip, *but not both*! There are two purposes for the ring.

1. To keep the stopcock from falling out entirely. Unfortunately, the rubber rings are not aware of this, and the stopcock often falls out anyway.
2. To provide a sideways pressure, pulling the stopcock in, so that it will not leak. Names and addresses of individuals whose stopcocks could not possibly leak and did so anyway will be provided on request. So provide a little sideways pressure of your own.

When you grease a glass stopcock (and you must), do it very carefully so that the film of grease does *not* spread into the area of the stopper that comes in contact with any of your compound (Fig. 36).

The Teflon Stopcock

In wide use today, the **Teflon stopcock** (Fig. 37) requires no grease and will not freeze up! The glass surrounding the stopcock *is not ground glass and cannot be used in funnels that require ground glass stopcocks!* The

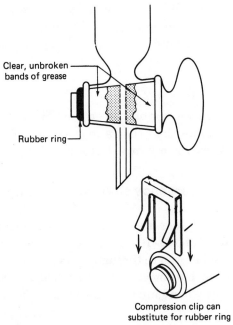

Fig. 36 The infamous glass stopcock.

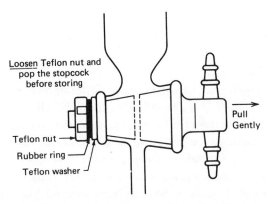

Fig. 37 Extreme closeup of the Teflon stopcock popping ritual.

Teflon stopcocks are infinitely easier to take care of. There is a Teflon washer, a rubber ring, and, finally, a Teflon nut, placed on the threads of the stopcock. This nut holds the whole thing on. Any leakage at this stopcock results from

1. A loose Teflon nut. Tighten it.
2. A missing Teflon washer or rubber ring. Have it replaced.
3. An attempt to place the wrong size or taper Teflon stopcock into the funnel. This is extremely rare. Get a new funnel.

Emergency stopcock warning!

Teflon may not stick, *but it sure can flow*! If the stopcock is extremely tight, the Teflon will bond itself to all the nooks and crannies in the glass in interesting ways. When you're through, always loosen the Teflon nut, and "pop the stopcock" by pulling on the handle. The stopcock should be loose enough to spin freely when spun with one finger—*then remember to tighten it again before you use it.*

It seems to me that I'm the only one who reads the little plastic bags that hold the stopcock parts. Right on the bags it shows that after the stopcock goes in, *the Teflon washer goes on the stem first,* followed by the rubber ring, and then the Teflon nut (Fig. 37). So why do I find most of these things put together incorrectly?

The Stem

The stem on a sep funnel can either be straight or have a ground glass joint on the end (Fig. 35). The ground glass joint fits the other jointware you may have and can be used that way as an **addition funnel** to add liquids or solutions into a setup (see "Addition and Reflux," p. 125). You can use this type of separatory funnel as a sep funnel. You can't, however, use the straight-stem separatory funnel as an addition funnel without some help; remember, straight glass tubes don't fit ground glass joints (see "The Adapter with Lots of Names," p. 21).

WASHING AND EXTRACTING VARIOUS THINGS

Now, getting back to extractions, there are really only four classes of compounds that are commonly handled in undergraduate extractions or washings.

1. *Strong acids.* The mineral acids and organic acids (e.g., benzoic acid). You usually *extract these into sodium bicarbonate solution or wash them with it.*
2. *Really weak acids.* Usually phenols, or substituted phenols. Here, you'd use a *sodium hydroxide solution for washing or extraction.* You need a *strong base* to work with these *weak acids.*
3. *Organic bases.* Any organic amine (aniline, triethylamine, etc.). As you use bases to work with acids, use a *dilute acid* (5–10% HCl, say) to extract or wash these bases.
4. *Neutral compounds.* All else, by these definitions (e.g., amides, ethers, alcohols, hydrocarbons).

HOW TO EXTRACT AND WASH WHAT

Here are some practical examples of washings and extractions, covering various types and mixtures and separations and broken down into the four classifications listed above.

1. *A strong organic acid.* Extract into sat'd (saturated) sodium bicarbonate solution.
   (*CAUTION! Foaming and fizzing and spitting and all sorts of carrying on.*)

   The weak base turns the strong acid into a salt, and the salt dissolves in the water–bicarbonate solution. Because of all the fizzing, you'll have to be very careful. Pressure can build up and blow the stopper out of the funnel. Invert the funnel. *Point the stem AWAY FROM EVERY-ONE up and toward the BACK OF THE HOOD*—and open the stopcock to vent or "burp" the funnel.

a. *To recover the acid,* add conc. (concentrated) HCl until the solution is acidic. Use pH or litmus paper to make sure. Yes, the solution really fizzes and bubbles. You should use a large beaker so material isn't thrown onto the floor if there's too much foam.

b. *To wash out the strong acid,* just throw the solution of bicarbonate away.

2. ***A weakly acidic organic acid.*** Extract into 10% NaOH–water solution. The strong base is needed to rip the protons out of weak acids (they don't want to give them up) and turn them into salts; then they'll go into the NaOH–water layer.

   a. *To recover the acid,* add conc. HCl until the solution of base is acid when tested with pH or litmus paper.

   b. *To wash out the weak acid,* just throw this NaOH–water solution away.

3. ***An organic base.*** Extract with 10% HCl–water solution. The strong acid turns the base into a salt. (This *turning the whatever into a salt that dissolves in the water solution* should be pretty familiar to you by now. Think about it. Then the salt goes into the water layer.

   a. *To recover the base,* add ammonium hydroxide to the water solution until the solution is *basic* to pH or litmus paper. *Note that this is the reverse of the treatment given to organic acids.*

   b. *To wash out an organic base, or any base,* wash as above and throw out the solution.

4. ***A neutral organic.*** If you've extracted strong acids first, then weak acids, then bases, there are only neutral compound(s) left. If possible, just remove the solvent that now contains *only your neutral compound.* If you have *more than one neutral compound,* you may want to extract one from the other(s). You'll have to find *two different immiscible organic liquids,* and *one liquid must dissolve ONLY the neutral organic compound you want!* A tall order. You must count on *one neutral organic compound being more soluble in one layer than in the other.* Usually the separation is *not clean—not complete.* And you have to do more work.

What's "more work"? That depends on the results of your extraction.

The Road to Recovery—Back-Extraction

11

I've mentioned **recovery** of the four types of extractable material, but that's not all the work you'll have to do to get the compounds in shape for further use.

1. If the recovered material is *soluble in the aqueous recovery solution,* you'll have to do a **back-extraction.**

    a. Find a solvent that dissolves your compound and is *not miscible in the aqueous recovery solution.* This solvent should boil at a low temperature (<100°C), since you will have to remove it. Ethyl ether is a common choice.
        (*HAZARD! Very flammable.*)
    b. *Then you extract your compound BACK FROM THE AQUEOUS RECOVERY SOLUTION into this organic solvent.*
    c. Dry this organic solution with a drying agent (see Chapter 23, "Drying Agents").
    d. Now you can remove the organic solvent. Either distill the mixture or evaporate it, perhaps on a steam bath. All this is done away from flames and in a hood.

    When you're through removing the solvent, if your product is not pure, clean it up. If your product is a liquid, you might distill it; if a solid, you might recrystallize it. Make sure it is clean.
2. If the recovered material is *insoluble in the aqueous recovery solution* and *it is a solid,* collect the crystals on a Büchner funnel (see Fig. 28). If they are *not pure,* you should recrystallize them.
3. If the recovered material is *insoluble in the aqueous recovery solution* and *it is a liquid,* you can use your separatory funnel to *separate the aqueous recovery solution from your liquid product. Then dry your liquid product and distill it if it is not clean. Or you might just do a back-extraction* as above. This has the added advantage of getting out the small amount of liquid product that dissolves in the aqueous recovery solution and increases your yield. Remember to dry the back-extracted solution before you remove the organic solvent. Then distill your liquid compound if it is not clean.

A SAMPLE EXTRACTION

I think the only way I can bring this out is to use a typical example. This may ruin a few lab quizzes, but if it helps, it helps.

Say you have to separate a mixture of *benzoic acid (1)*, *phenol (2)*, *p-toluidine (4-methylanaline) (3)*, and *anisole (methoxybenzene) (4)* by extraction. The numbers refer to class of compound, as listed above. We're assuming that none of the compounds reacts with any of the others and that you know that we're using all four types as indicated. Phenol and 4-methylanaline are corrosive toxic poisons and if you get near these compounds in lab, *be very careful*. When they are used as an example on these pages, however, you are quite safe. Here's a sequence of tactics.

1. Dissolve the mixture in ether. Ether is insoluble in the water solutions you will extract into. Ether happens to dissolve all four compounds. Aren't you lucky? You bet! It takes lots of hard work to come up with the "typical student example."
2. Extract the ether solution with 10% HCl. This converts *only* compound 3, the basic *p*-toluidine, into the hydrochloride salt, which dissolves in the 10% HCl layer. You have just *extracted a base with an acid solution*. Save this solution for later.
3. Now extract the ether solution with sat'd sodium bicarbonate solution. *Careful!* Boy will this fizz! Remember to swirl the contents and release the pressure. The weak base converts *only* compound 1, the *benzoic acid*, to a salt, which dissolves in the sat'd bicarbonate solution. Save this for later.
4. Now extract the ether solution with the 10% NaOH solution. This converts the compound 2, *weak acid*, phenol, to a salt, which dissolves in the 10% NaOH layer. Save this for later. *If you do this step before step 3*, that is, extract with *10% NaOH solution before the sodium bicarbonate* solution, *both the weak acid*, phenol, *and the strong acid*, benzoic acid, *will be pulled out into the sodium hydroxide*. Ha ha. This is the usual kicker they put in lab quizzes, and people always forget it.
5. The only thing left is the neutral organic compound dissolved in the ether. Just drain this into a flask.

So, now we have four flasks with four solutions with one component in each. *They are separated.* You may ask, "How do we get these back?"

1. *The basic compound (3).* Add ammonium hydroxide until the solution turns basic (test with litmus or pH paper). The *p*-toluidine, an organic base (3), is regenerated.
2. *The strong acid or the weak acid (1, 2).* A bonus. Add dilute HCl until the solution turns acidic to an indicator paper. Do it to the other solution. Both acids will be regenerated.
3. *The neutral compound (4).* It's in the ether. If you evaporate the ether (*no flames!*), the compound should come back.

Now, when you recover these compounds, sometimes they don't come back in such good shape. You will have to do more work.

1. Addition of HCl to the benzoic acid extract will produce huge amounts of white crystals. Get out the Büchner funnel and have a field day! Collect all you want. But they won't be in the best of shape. Recrystallize them.
    *(**NOTE:** This compound is insoluble in the aqueous recovery solution.)*

2. The phenol extract is a different matter. You see, *phenol is soluble in water,* and it doesn't come back well at all. So, get some fresh ether, extract the phenol from the HCl solution to the ether, and evaporate the ether. Sounds crazy, no? No. Remember, I called this a **back-extraction** and you'll have to do this more often than you would like to believe.
    *(**NOTE:** Phenol is soluble in the aqueous recovery solution.)*

3. The *p*-toluidine should return after the addition of ammonium hydroxide. Recrystallize it from ethanol so it also looks respectable again.
4. The neutral anisole crystals that come back after evaporating the ether (*no flames!*) will probably be contaminated with a little bit of all the other compounds that started out in the ether. You must collect and recrystallize this solid as well.

You may or may not have to do all this with the other solutions, or with any other solution you ever extract in your life. You must choose. Art over science. As confusing as this is, I have simplified it a lot. Usually you have to extract these solutions more than once, and the separation is not as clean as you'd like. Not 100%, but pretty good. If you are still confused, see your instructor.

PERFORMING AN EXTRACTION OR WASHING

1. Suspend a sep funnel in an iron ring.
2. Remove the stopper.
3. *Make sure the stopcock is closed*! You don't really want to scrape your product off the benchtop.
4. Add the solution to be extracted or washed. Less than half full, please. Add the extraction or washing solvent. An equal volume is usually enough. The funnel is funnel shaped, and the equal volumes won't look equal.
5. Replace the stopper.
6. Remove the sep funnel from the iron ring. Hold stopper and stopcock tightly. Pressure may build up during the next step and blow your product out onto the floor.
7. Invert the sep funnel (Fig. 38).

**Point the stem up away from everyone—
up into the back of a hood if at all possible!**

Make *sure* the liquid has drained down away from the stopcock, then *slowly* open the stopcock. You may hear a woosh, possibly, a pffffft, as the pressure is released. This is due to the high vapor pressure of some solvents or to a gas evolved from a reaction during the mixing. This can cause big trouble when you are told to neutralize acid by washing with sodium carbonate or sodium bicarbonate solutions.

8. *CLOSE THE STOPCOCK!*
9. Shake the funnel gently, invert it, and open the stopcock again.
10. Repeat steps 8 and 9 until no more gas escapes.

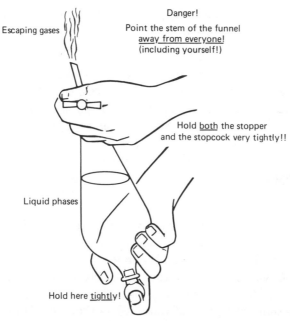

Escaping gases

Danger!
Point the stem of the funnel
<u>away from everyone!</u>
(including yourself!)

Hold <u>both</u> the stopper
and the stopcock very tightly!!

Liquid phases

Hold here <u>tightly</u>!

Fig. 38   Holding a sep funnel so as not to get stuff all over.

11. If you see that you might get an **emulsion**—*a fog of particles*—with this gentle inversion, *do NOT shake the funnel vigorously.* You might have to continue the rocking and inverting motions 30 to 100 times, as needed, to get a separation. Check with your instructor, and read about breaking up emulsions (see "Extraction Hints," below). Otherwise, shake the funnel vigorously about 10 times to get good distribution of the solvents and solutes. Really shake it.

12. Put the sep funnel back in the iron ring.

13. *Remove the glass stopper*. Otherwise the funnel won't drain and you'll waste your time just standing there.

14. Open the stopcock, and let the bottom layer drain off into a flask.

15. Close the stopcock, swirl the funnel gently, then wait to see if any more of the bottom layer forms. If so, collect it. If not, assume you got it all in the flask.

16. Let the remaining layer out into another flask.

To extract any layer again, return that layer to the sep funnel, *add fresh extraction or washing solvent,* and repeat this procedure starting from step 5.

Never, never, never, never,
ever throw away any layer, until you are absolutely sure you'll
never need it again. Not very much of your product can be re-
covered from the sink trap!

EXTRACTION HINTS

1. Several smaller washings or extractions are better than one big one.
2. Extracting or washing a layer *twice,* perhaps thrice, is usually enough. Diminishing returns set in after that.
3. Sometimes you'll have to find out which layer is **the water layer**. This is so simple, it confounds everyone. Add 2–4 drops of each layer to a test tube containing 1 ml of water. Shake the tube. If the stuff *doesn't dissolve* in the water, it's *not* an aqueous (water) layer. The stuff may sink to the bottom, float on the top, do *both,* or even *turn the water cloudy*! It will *not,* however, dissolve.
4. If *only the top layer* is being extracted or washed, *it does not have to be removed from the funnel,* ever. Just drain off the bottom layer, then add more *fresh* extraction or washing solvent. Ask your instructor about this.
5. You *can* combine the extracts of a multiple extraction *if they have the same material in them.*
6. If you have to wash your organic compound with water, and the organic is *slightly soluble in water,* try washing with *saturated salt solution.* The theory is that if all that salt is dissolved in the water, what room is there for your organic product? This point is a favorite of quizmakers and should be remembered. It's the same thing that happens when you add salt to reduce the solubility of your compound during a crystallization (see "Salting Out," p. 61).
7. If you get an **emulsion,** you have not two distinct layers but a kind of a *fog of particles*. Sometimes you can break up the charge on the

suspended droplets by adding a little salt or some acid or base. Careful with the acids and bases though. They can react with your product and destroy it.

8. If you decide to add salt to a sep funnel, don't add so much that it clogs the stopcock! For the same reason, keep drying agents out of sep funnels.

9. Sometimes some material comes out or will not dissolve in the two liquid layers and hangs in there in the **interface.** It may be that there's not enough liquid to dissolve this material. One cure is to *add more fresh solvent of one layer or the other*. The solid may dissolve. If there's no room to add more, you may have to remove *both* (yes, both) layers from the funnel and try to dissolve this solid in either of the solvents. It can be confusing. If the material does *not* redissolve, it is a new compound and should be saved for analysis. You should see your instructor for that one.

And Now— Boiling Stones

12

All you want to do is start a distillation. Instructor walks up and says,
"Use a boiling stone or it'll bump."
"But I'm only gonna ..."
"Use a boiling stone or it'll bump."
"It's started already and ..."
"Use a boiling stone or it'll bump."
"I'm not gonna go and. ..."

Suddenly—WOOSH! Product. All over the bench! Instant failure! Next time you put a boiling stone in *before* you start. No bumping. But your instructor won't let you forget the time you did it your way.

Don't let this happen to you. Use a **brand new boiling stone** every time you have to boil a liquid. A close comparison between a boiling stone and the inner walls of a typical glass vessel reveals thousands of tiny nucleating points on the stone where vaporization can take place, in contrast to the smooth glass surface that can hide unsightly hot spots and lead to BUMPING, a massive instantaneous vaporization that will throw your product all over.

CAUTION!   Introduction of a boiling stone into hot liquid may result in instant vaporization and loss of product. Remove the heat source, swirl the liquid to remove hot spots, then add the boiling stone.

Used as directed, the boiling stone will relieve minor hot spots and prevent loss of product through bumping. So remember ... *whenever you boil, wherever you boil,*

ALWAYS USE A FRESH BOILING STONE!

Don't be the last on your bench to get this miracle of modern science made exclusively from nature's most common elements.

Sources of Heat

13

Many times you'll have to heat something. Don't just reach for the Bunsen burner. That flame you start may be your own. There are alternate sources you should think of *first*.

THE STEAM BATH

If one of the components boils below 70°C, and you use a *Bunsen burner*, you may have a hard time putting out the fire. Use a **steam bath**!

1. Find a steam tap. It's like a water tap, only it dispenses steam. *(CAUTION! You can get burned.)*
2. Connect tubing to the tap *now*. It's going to get awfully hot in use. Make sure you've connected a piece that'll be long enough to reach your steam bath.
3. Don't connect this tube to the steam bath yet! Just put it into a sink. Because steam lines are usually full of water from condensed steam, *drain the lines first;* otherwise you'll waterlog your steam bath.
4. *Caution!* Slowly open the steam tap. You'll probably hear bonking and clanging as steam enters the line. Water will come out. It'll get hotter and may start to spit.
5. Wait until the line is mostly clear of water, then turn off the steam tap. *Wait for the tubing to cool.*
6. Slowly, carefully, and cautiously, *making sure the tube is not hot,* connect the tube to the inlet of the steam bath. This is the *uppermost* connection on the steam bath.
7. Connect another tube to the outlet of the steam bath—the *lower* connection—and to a drain. Any water that condenses in the bath while you're using it will drain out.

Usually, steam baths have concentric rings as covers. You can control the "size" of the bath by adding or removing rings.

Never do this after you've started the steam. You will get burned!

And don't forget—round-bottom flasks should be about halfway in the bath with steam rising up all around the flask, not coming out the sides of the bath or any other place (Fig. 39).

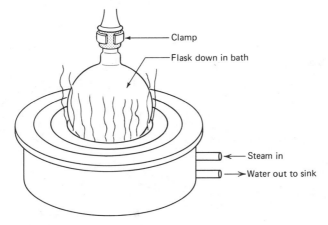

Fig. 39 The steam bath in use.

THE BUNSEN BURNER (Fig. 40)

The first time you get the urge to take out a Bunsen burner and light it up, *don't.* You may blow yourself up. Please check with your instructor to see if you even need a burner. Once you find out that you *can* use a burner, assume that the person who used it last didn't know much about burners, and take some precautions so as not to burn your eyebrows off.

Now Bunsen burners are not the only kind. There are **Tirrill burners** and **Meker burners** as well. Some are more fancy than others, but they work pretty much the same. So when I say **burner** anywhere in the text it could be any of them.

1. Find the **needle valve.** This is at the base of the burner. Turn it fully clockwise (inward) to stop the flow of gas completely. If your burner doesn't have a needle valve, it's a traditional Bunsen burner, and the gas flow is regulated at the bench stopcock (Fig. 40). This can be dangerous, especially if you have to reach over your apparatus and burner to turn off the gas. Try to get a different model.
2. There is a **movable collar** at the base of the burner which controls air flow. For now, see that *all* the holes are closed (*i.e.,* no air gets in).

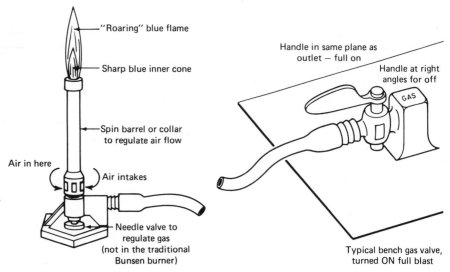

"Roaring" blue flame

Sharp blue inner cone

Spin barrel or collar
to regulate air flow

Air in here

Air intakes

Needle valve to
regulate gas
(not in the traditional
Bunsen burner)

Handle in same plane as
outlet — full on

Handle at right
angles for off

GAS

Typical bench gas valve,
turned ON full blast

Fig. 40 More than you may care to know about burners.

3. Connect the burner to the bench stopcock by some tubing, and turn the bench valve *full on*. The bench valve handle should be parallel to the outlet (Fig. 40).

4. Now, *slowly* open the needle valve. You may be just able to hear some gas escaping. Light the burner. *Mind your face!* Don't look down at the burner as you open the valve.

5. You'll get a wavy yellow flame, something you don't really want. But at least it'll light. Now open the air collar a little. The yellow disappears; a blue flame forms. This is what you want.

6. Now, adjust the needle valve and collar (the adjustments play off each other) for a steady blue flame.

Burner Hints

1. When you set up a distillation or reflux, don't waste a lot of time *raising* and *lowering* the entire setup so the burner will fit. This is nonsense. Move the burner! Tilt it! See Fig. 41. If you leave the burner motionless, under the flask, you may scorch the compound and your precious product may become a "dark intractable material."

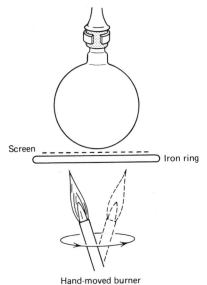

Screen

Iron ring

Hand-moved burner

Fig. 41 Don't raise the flask, lower the burner.

2. Placing a wire gauze between the flame and the flask spreads out the heat evenly. Even so, the burner may have to be moved around. Hot spots can cause star cracks to appear in the flask (see Fig. 6).

3. *Never* place the flask in the ring without a screen (Fig. 42). The iron ring heats up faster than the flask, and the flask cracks in the nicest line around it you've ever seen. The bottom falls off, and the material is all over your shoes.

THE HEATING MANTLE

A very nice source of heat, the heating mantle takes some special equipment and finesse.

1. ***Variable voltage transformer.*** The transformer takes the quite lethal 120 V from the wall socket and can change it to an equally dangerous 0 to 120 V, depending on the setting on the dial. Unlike temperature settings on a Mel-Temp, on a transformer 0 means 0 V, 20 means 20 V, and so on. I like to start at 0 V and work my way up.

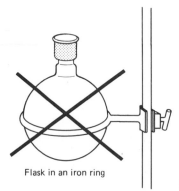

Flask in an iron ring

Fig. 42 Flask in an iron ring.

Depending on how much heat you want, values from 40 to 70 seem to be good places to start. Also, you'll need a cord that can plug into both the transformer and the heating mantle.

2. ***The heating mantle itself.*** An electric heater wrapped in fiberglass insulation and cloth that looks vaguely like a catcher's mitt.

3. ***Things not to do.***

   a. *Don't ever plug the mantle directly into the wall socket!* I know, the curved prongs on the mantle connection won't fit, but the straight prongs on the adapter cord will. Always use a variable voltage transformer, and start with the transformer set to zero.

   b. *Don't use too small a mantle.* The only cure for this is to *get one that fits properly.* The poor contact between the mantle and the glass doesn't transfer heat readily and the mantle burns out.

   c. *Don't use too large a mantle.* The only good cure for this is to *get one that fits properly.* An acceptable fix is to *fill the mantle with sand, after the flask is in,* but before you turn the voltage on. Otherwise, the mantle will burn out.

HINT. When you set up a heating mantle to heat any flask, usually for **distillation** or **reflux,** put the mantle on an iron ring, and keep it clamped a few inches above the desktop (Fig. 43). Then clamp the flask *at the neck,* in case you have to remove the heat quickly. You can just unscrew the lower clamp and drop the mantle and iron ring.

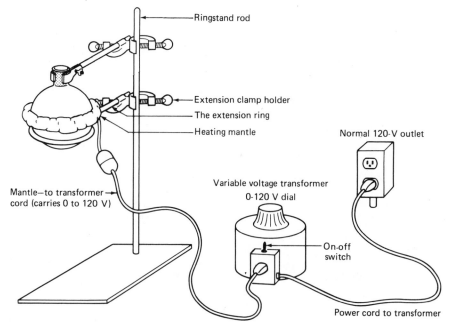

Labels in figure:
- Ringstand rod
- Extension clamp holder
- The extension ring
- Heating mantle
- Normal 120-V outlet
- Mantle—to transformer cord (carries 0 to 120 V)
- Variable voltage transformer 0-120 V dial
- On-off switch
- Power cord to transformer

13

Fig. 43 R.B. flask and mantle ready to go.

Clamps
and
Clamping

14

Unfortunately, glass apparatus needs to be held in place with more than spit and bailing wire. In fact you would do well to use clamps. Life would be simple if there were just one type of fastener, but that's not the case.

1. ***The simple buret clamp*** (Fig. 44). Though popular in other chem labs, the simple buret clamp just doesn't cut it for organic lab. The clamp is too short, and adjusting angles with the "locknut" (by loosening the locknut, swiveling the clamp jaws to the correct angle, and tightening the locknut against the *back* stop, away from the jaws) is not a great deal of fun. If you're not careful, the jaws'll slip right around, and all the chemicals in your flask will fall out.

2. ***The simple extension clamp and clamp fastener*** (Fig. 45). This two-piece beast is the second best clamp going. It is much longer (approx. 12 in.), so you can easily get to complex setups. By loosening the **clamp holder thumbscrew,** the clamp can be pulled out, or pushed back, or rotated to any angle. By loosening the **ringstand thumbscrew,** the clamp, along with the clamp holder, can move up and down.

3. ***The three-fingered extension clamp*** (Fig. 46). Truly the Cadillac of

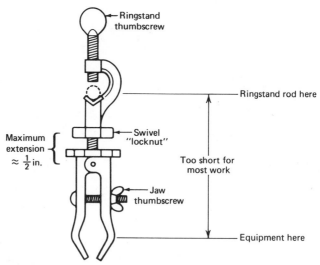

Fig. 44 The "barely adequate for organic lab" buret clamp.

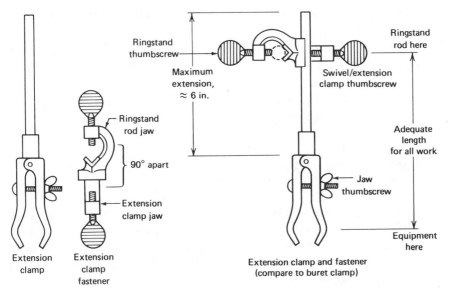

Ringstand
thumbscrew

Ringstand
rod here

Maximum
extension,
≈ 6 in.

Swivel/extension
clamp thumbscrew

Ringstand
rod jaw

90° apart

Extension
clamp jaw

Adequate
length
for all work

Jaw
thumbscrew

Equipment
here

Extension
clamp

Extension
clamp
fastener

Extension clamp and fastener
(compare to buret clamp)

Fig. 45 The extension clamp and clamp fastener.

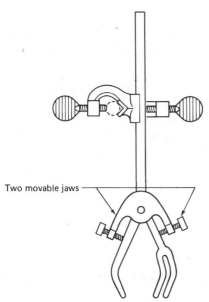

Two movable jaws

Fig. 46 The three-fingered clamp with clamp fastener.

clamps with a price to match. They usually try to confuse you with *two thumbscrews* to tighten, unlike the regular extension clamp. This gives a bit more flexibility, at the cost of a slightly more complicated way of setting up. You can make life simple by opening the *two-prong bottom jaw to a 10 to 20° angle from the horizontal and treating that jaw as fixed.* This will save a lot of wear and tear when you set equipment up, but you can *always move the bottom jaw* if you have to.

CLAMPING A DISTILLATION SETUP

You'll have to clamp many things in your life as a chemist, and one of the more frustrating setups to clamp is the **simple distillation** (see Chapter 15, "Distillation"). If you can set this up, you probably will be able to clamp other common setups without much trouble. Here's how to go about setting up the simple distillation.

1. OK, get a **ringstand** and an **extension clamp and clamp fastener** and put them all together. What heat source? A Bunsen burner, and you'll need more room than with a heating mantle (see Chapter 13, "Sources of Heat"). In any case, you don't know where the receiving flask will show up; and then you might have to readjust the entire setup. Yes, you should have read the experiment before so you'd know about the heating mantles.
2. Clamp the flask (around the neck) a few inches up the ringstand (Fig. 47). We *are* using heating mantles and you'll need the room underneath to drop the mantle in case it gets too hot. That's why the flask is *clamped at the neck. Yes. That's where the flask is ALWAYS clamped, no matter what heat source.* So it doesn't fall when the mantle comes down. What holds the mantle? Extension ring and clamp fastener.
3. Remember, always set these up from left to right—*distilling flask first!*
4. Add the three-way adapter now (Fig. 48). Thermometer and thermometer adapter comes later.
5. Now add the condenser. Get another ringstand, extension clamp, and clamp fastener. There. Estimate the angle and height the clamp will

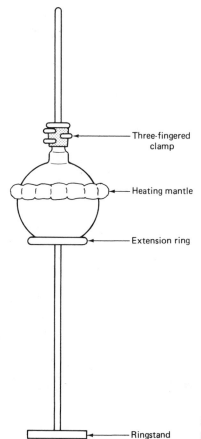

Fig. 47 Flask and heating mantle out on a ringstand.

be at when the setup is clamped. Try setting the two-pronged jaw at about a 30° angle to the extension rod and call that the fixed jaw. Now turn the clamp so that the two-prong "fixed" jaw is at the bottom. Now, this two-pronged jaw of the clamp acts as a cradle for the condenser. Since tightening the top jaw won't move the bottom jaw, there won't be too many problems.

6. Place the two-prong clamp jaw in line with the condenser. Attach the condenser to the three-way adapter (Fig. 49). Hold everything! Sure. OK, *loosen extension clamp holder thumbscrew,* turn clamp to correct angle, and tighten. Now the height—up just a bit—good! The lower

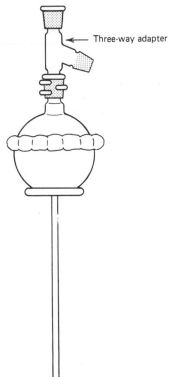

Three-way adapter

Fig. 48 Clamps and flask and three-way adapter.

"fixed" jaw cradles the condenser. *Tighten the ringstand thumbscrew (Arrgh!). Clamps tend to move up ever so slightly as you tighten the fastener on the ringstand.*

7. Bottom jaw supports condenser—check. Joint at the three-way adapter/condenser OK? Good. Tighten wing nut and *bring single-prong jaw down* onto condenser (Fig. 50). Not too tight.

8. Back from the stockroom again. Having put the vacuum adapter on the end of the condenser, expecting it to stay there by magic, you'll be more careful with the new one.

9. Do the third ringstand–extension clamp–clamp fastener setup. It's

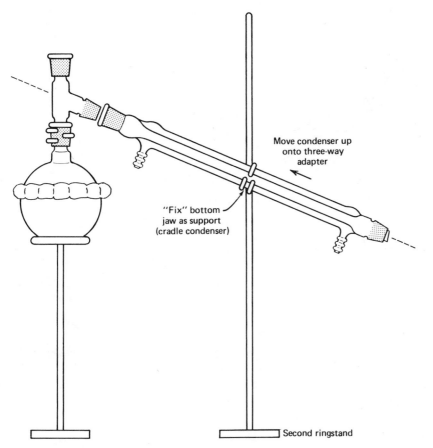

Move condenser up
onto three-way
adapter

"Fix" bottom
jaw as support
(cradle condenser)

Second ringstand

14

Fig. 49 Clamping the condenser without arthritic joints.

handy to think of the extension clamp and clamp fastener as a single unit. Clamp receiving flask in place. Put vacuum adapter in the flask now. Adjust. There! Got it.

10. All the clamps set up, all the joints tight—now where is that thermometer adapter?

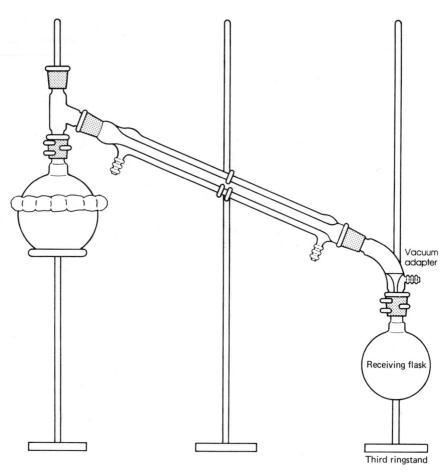

Fig. 50 Correctly clamping the vacuum adapter.

Distillation

The separation or purification of liquids by vaporization and condensation is a very important step in one of man's oldest professions. The word "still" lives on as a tribute to the importance of organic chemistry. The important points are

1. **Vaporization.** Turning a liquid to a vapor.
2. **Condensation.** Turning a vapor to a liquid.

Remember these. They show up on quizzes.

But when do I use distillation? That is a very good question. Use the guidelines below to pick your special situation, and turn to that section. But you *should* read *all* the sections, anyway.

1. **Class 1: Simple distillation.** Separating liquids boiling BELOW 150°C at one atmosphere (1 atm) from

    a. Nonvolatile impurities.
    b. Another liquid boiling at least 25°C higher than the first. The liquids should dissolve in each other.

2. **Class 2: Vacuum distillation.** Separating liquids boiling ABOVE 150°C at 1 atm from

    a. Nonvolatile impurities.
    b. Another liquid boiling at least 25°C higher than the first. They should dissolve in one another.

3. **Class 3: Fractional distillation.** Separating liquid mixtures, soluble in each other, that boil at less than 25°C from each other at 1 atm.
4. **Class 4: Steam distillation.** Isolating tars, oils, and other liquid compounds *insoluble,* or slightly soluble, *in water at all temperatures.* Usually natural products are steam distilled. They do *not* have to be liquids at room temperatures (*e.g.,* caffeine, a solid, can be isolated from green tea).

Remember, these are guides. If your compound boils at 150.0001°C don't scream that you MUST do a vacuum distillation or both you and your product will die. I expect you to have some judgment and to pay attention to your instructor's specific directions.

DISTILLATION NOTES

1. *EXCEPT for class 4,* steam distillation, two liquids that are to be separated must dissolve in each other. If they did not, they would form separable layers, which you could separate in a separatory funnel (see Chapter 11, "Extraction").
2. Impurities can be either **soluble** or **insoluble**. For example, the material that gives cheap wine its unique bouquet is soluble in the alcohol. If you distill cheap wine, you get clear grain alcohol separated from the "impurities," which are left behind in the distilling flask.

CLASS 1: SIMPLE DISTILLATION (Fig. 51)

For separation of liquids boiling below 150°C at 1 atm from

15

1. Nonvolatile impurities.
2. Another liquid boiling 25°C higher than the first liquid. *They must dissolve in each other.*

Sources of Heat

If one of the components boils below 70°C, and you use a Bunsen burner, you may have a hard time putting out the fire. Use a steam bath or a heating mantle. Different distillations will require different handling (see Chapter 13, "Sources of Heat"). All the distillations always require heating, so the **sources of heat** chapter is really closely tied to this section. This goes for enlightenment on the use of **boiling stones** and **clamps** as well (see Chapters 12 and 14, respectively).

The Three-Way Adapter

If there is any *one place* your setup will fall apart, here it is (Fig. 52). When you set up the jointware, it is important that you have all the joints *line up*.

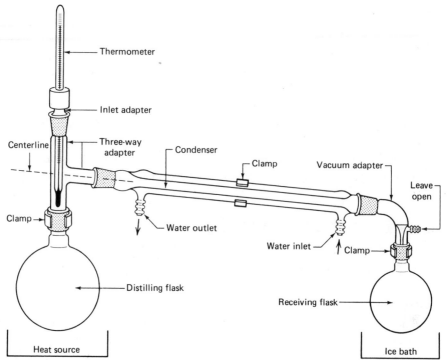

Fig. 51 A complete, entire simple distillation setup.

This is tricky, since, as you push one joint together, another pops right out. If you're not sure, call your instructor. Let him inspect your work. Remember,

All joints must be tight!

The Distilling Flask

Fill the distilling flask with the liquid you want to distill. You can remove the thermometer and thermometer adapter, fill the flask using a funnel, then put the thermometer and its adapter back in place.

If you're doing a **fractional distillation** with a **column** (a class 3 distillation, see p. 113), you should've filled the flask *before* clamping the setup. (Don't ever pour your mixture down a column. That'll contaminate

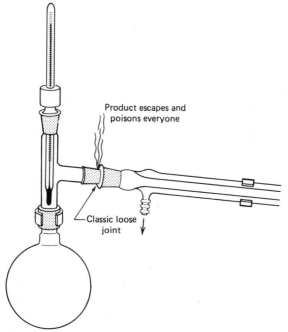

Fig. 52 The "commonly camouflaged until it's too late" open joint.

everything!) You'll just have to disassemble some of the setup, fill the flask, reassemble what you've taken down, and pray that you haven't knocked all the other joints out of line.

Don't fill the distilling flask more than half full. Put in a boiling stone if you haven't already. These porous little rocks promote bubbling and keep the liquid from superheating and flying out of the flask. This flying around is called **bumping**. NEVER drop a boiling stone into hot liquid or you may be rewarded by having your body soaked in the hot liquid as it foams out at you.

Make sure all the joints in your setup are tight. Start up the heat S–L–O–W–L–Y until gentle boiling begins and liquid starts to drop into the receiving flask at the rate of about 10 drops per minute. *This is important.* If nothing comes over, you're not distilling, merely wasting time. You may have to turn up the heat to keep material coming over.

The Thermometer Adapter

Read all about it. Ways of having fun with thermometer adapters have been detailed (see text accompanying Fig. 8).

The Ubiquitous Clamp

A word about clamps. *Use!* They can save you $68.25 in busted glassware (see Chapter 14, "Clamps and Clamping").

The Thermometer

Make sure the thermometer bulb is *below the sidearm of the three-way adapter.* If you don't have liquid droplets condensing on the thermometer bulb, the temperature you read is *nonsense.* Keep a record of the temperature of the liquid or liquids that are distilling. It's a check on the purity. Liquid collected over 2°C range is fairly pure. Note the similarity of this range with that of the *melting point* of a pure compound (see Chapter 9, "The Melting Point Experiment").

The Condenser

Always keep cold water running through the condenser, enough so that *at least the lower half is cold* to the touch. Remember that water should go *in the bottom* and *out of the top* (Fig. 51). Also, the water pressure in the lab may change from time to time and usually goes up at night, since little water is used then. So, if you are going to let condenser cooling water run overnight, tie the tubing on at the condenser and the water faucet with wire or something. And if you don't want to flood out the lab, see that the outlet hose can't flop out of the sink.

The Vacuum Adapter

It is important that the tubing connector remain *open to the air,* or the entire apparatus will, quite simply, explode.

WARNING: Do not just stick the vacuum adapter on the end of the condenser and hope that it will not fall off and break.

This is foolish. I have no sympathy for people who will not use clamps to save their own breakage fee. They deserve it.

The Receiving Flask

The receiving flask should be large enough to collect what you want. You may need several, and they may have to be changed during the distillation. Standard practice is to have ONE flask ready for what you are going to throw away and others ready to save the stuff that you want to save.

The Ice Bath

Why everyone insists on loading up a bucket with ice and trying to force a flask into this mess, I'll never know. How much cooling do you think you're going to get with a few small areas of the flask barely touching ice? Get a suitable receptacle—a large beaker, enameled pan, or whatever. *It should not leak.* Put it under the flask. Put some water in it. *Now add ice.* Stir. Serves four.

Ice bath really means ice-water bath

THE DISTILLATION EXAMPLE

Say you place 50 ml of liquid A (B.P. 50°C) and 50 ml of liquid B (B.P. 100°C) in a 250-ml R.B. flask. You drop in a boiling stone, fit the flask in a distillation setup, and turn on the heat. Bubbling starts and soon droplets form on the thermometer bulb. The temperature shoots up *from room temperature to about 35°C,* and a liquid condenses and drips into the receiver. That's bad. The temperature should be close to 50°C. This low-boiling material is the **fore–run** of a distillation, and you won't want to keep it.

Keep letting liquid come over until the temperature stabilizes at about 49°C. Quick! Change receiving flasks NOW!

The new receiving flask is on the condenser, and the temperature is about 49°C. GOOD. Liquid comes over and you heat to get a rate of about 10 drops per minute collected in the receiver. As you distill, the temperature slowly increases to maybe 51°C then starts moving up rapidly.

Here you stop the distillation and change the receiver. Now in one receiver you have *a pure liquid, B.P. 49–51°C*. Note this **boiling range.** It is just as good a test of purity as a melting point is for solids (see Chapter 9 again).

Always report a boiling point for liquids as routinely as you report melting points for solids. The boiling point is actually a **boiling range** and should be reported as such:

"B.P. 49–51°C"

If you now put on a new receiver, and start heating again, you may discover *more material coming over at 50°C!* Find that strange? Not so. All it means is that you were distilling too rapidly, and some of the low-boiling material was left behind. It is very difficult to avoid this situation. Sometimes it is best to ignore it, unless a yield is very important. You can combine this "new" 50°C fraction with the other good fraction.

For liquid B, boiling at 100°C, merely substitute some different boiling points and go over the same story.

THE DISTILLATION MISTAKE

OK, you set all this stuff up to do a distillation. Everything's going fine. Clamps in the right place, no arthritic-joints, even the vacuum adapter is clamped on, and the thermometer is at the right height. There's a bright golden haze on the meadow, and everything's going your way. So you begin to boil the liquid. You even remembered the boiling stone. Boiling starts slowly, then goes more rapidly. You think, "This is *it!*" Read that temperature, now. Into the notebook: "The mixture started boiling at 26°C."

You are dead wrong.

What happened? The *SAME* thing that happens *ALL* the time.

Is there liquid condensing on the thermometer bulb??
NO!

So, congratulations, you've just recorded the room temperature. There are days when over half the class will report distillation temperatures as "Hey I see it start boiling now" temperatures. Don't participate. Just keep watching as the liquid boils. Soon, droplets *will* condense on the thermometer bulb. The temperature will go up quickly, then *stabilize*. NOW read the temperature. That's the boiling point.

CLASS 2: VACUUM DISTILLATION

For separation of liquids boiling above 150°C at 1 atm from

15

1. Nonvolatile impurities.
2. Another liquid boiling 25°C higher than the first liquid.

They must dissolve in each other.
This is like the **simple distillation** with the changes shown (Fig. 53).
Why vacuum distill? If the substances boil at high temperatures at 1 atm, they may decompose when heated. Putting a vacuum over the liquid makes the liquid boil at a lower temperature. With the pressure reduced, there are fewer molecules in the way of the liquid you are distilling. Since the molecules require less energy to leave the surface of the liquid, *you can distill at a lower temperature,* and *your compound doesn't decompose.*

Vacuum Distillation Notes

1. Read *ALL* the notes on class 1.
2. The thermometer can be replaced by a **gas inlet tube**. It has a long, fine capillary at one end (Fig. 53). This is to help stop the *extremely bad bumping* that goes along with vacuum distillations. The fine stream of

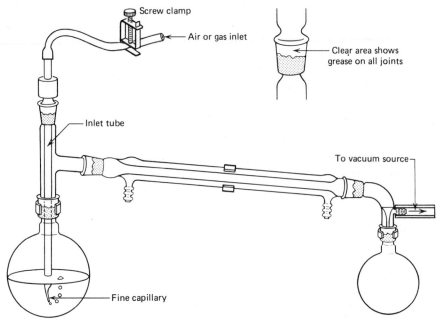

Fig. 53 A vacuum distillation setup.

bubbles through the liquid produces the same results as a boiling stone. Boiling stones are useless, since all the adsorbed air is whisked away by the vacuum and the nucleating cavities plug up with liquid. The fine capillary does not let in a lot of air, so we are doing a vacuum distillation anyway. Would you be happier if I called it a **reduced-pressure** distillation? An inert gas (nitrogen?) may be let in if the compounds decompose in air.

3. If you can get a **magnetic stirrer** and **magnetic stirring bar** you won't have to use the gas inlet tube approach. Put a magnetic stirring bar in the flask with the material you want to vacuum distill. Use a **heating mantle** to heat the flask, and put the magnetic stirrer under the mantle. When you turn the stirrer on, a magnet in the stirrer spins, and the stirring bar (a Teflon-coated magnet) spins. Admittedly, stirring through a heating mantle is not easy, but it can be done. *Stirring the liquid also stops the bumping.*

Remember, first the stirring, then the vacuum, THEN the heat—or WOOSH! Got it?

4. Control of heating is *extremely critical*. I don't know how to shout this loudly enough on paper. Always apply the vacuum first and watch the setup for awhile. Air dissolved or trapped in your sample or a highly volatile leftover (maybe ethyl ether from a previous extraction) can come flying out of the flask *without any heat*. If you heated such a setup a bit and then applied the vacuum, your sample would blow all over, possibly right into the receiving flask. Wait for the contents of the distilling flask to calm down before you start the distillation.

5. If you know you have low-boiling material in your compound, think about distilling it at atmospheric pressure first. If, say, half the liquid you want to vacuum distill is ethyl ether from an extraction, consider doing a simple distillation to get rid of the ether. Then the ether (or any other low-boiling compound) won't be around to cause trouble during the vacuum distillation. If you distill first at 1 atm, *let the flask cool BEFORE you apply the vacuum*. Otherwise your compound will fly all over and probably will wind up, undistilled and impure, in your receiving flask.

6. *Grease all joints, no matter what* (see "Greasing the Joints," p. 24). Under vacuum, it is easy for any material to work its way into the joints and turn into concrete, and the joints will never, ever come apart again.

7. The **vacuum adapter** is connected to a **vacuum source,** either a **vacuum pump** or a **water aspirator**. Real live vacuum pumps are expensive and rare and not usually found in the undergraduate organic laboratory. If you can get to use one, that's excellent. See your instructor for the details. The **water aspirator** is used lots, so read up on it (p. 56).

8. During a vacuum distillation, it is not unusual to collect a *pure compound over a 10–20°C temperature range*. If you don't believe it, you haven't ever done a vacuum distillation. It has to do with pressure changes throughout the distillation because the setup is far from perfect. Although a vacuum distillation is not difficult, it requires

15

peace of mind, large quantities of patience, and a soundproof room to scream in so as not to disturb others.

9.   A **Claisen adapter** in the distilling flask allows temperature readings to be taken and can help stop your compound from splashing over into the distillation receiver (Fig. 54). Also, you could use a **three-neck flask** (Fig. 55). Think! And, of course, use some glassware too.

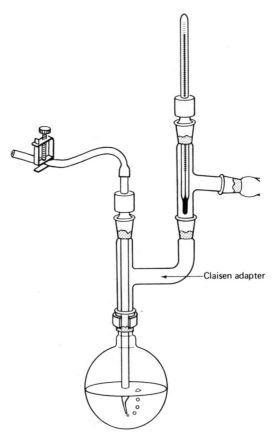

Fig. 54 A Claisen adapter so you can vacuum distill and take temperatures too.

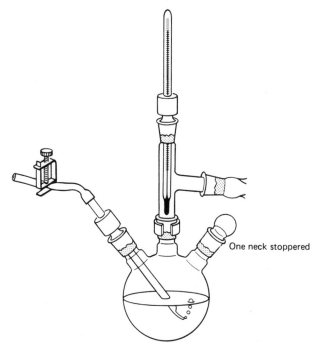

One neck stoppered

15

Fig. 55 Same multipurpose setup with a three-neck flask.

CLASS 3: FRACTIONAL DISTILLATION

For separation of liquids, soluble in each other, that boil less than 25°C from each other, use fractional distillation. This is like simple distillation with the changes shown (Fig. 56).

Fractional distillation is used when the components to be separated boil *within 25°C of each other*. Each component is called a **fraction**. Clever where they get the name, eh? This temperature difference is not gospel. And don't expect terrific separations either. Let's just leave it at *close boiling points*. How close? That's hard to answer. Is an orange? That's easier to answer. If the experiment tells you to "fractionally distill," at least you'll be able to set it up right.

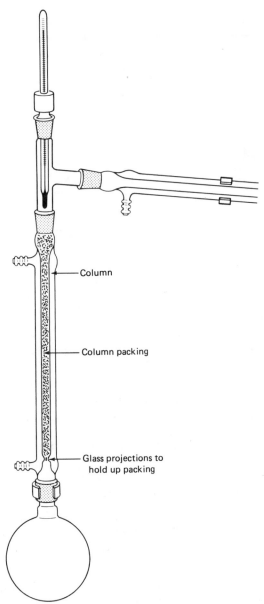

— Column

— Column packing

— Glass projections to
  hold up packing

Fig. 56 The fractional distillation setup.

How This Works

If one distillation is good, two are better. And fifty? Better still. So you have lots and lots of little, tiny distillations occurring on the surfaces of the **column packing,** which can be glass beads, glass helices, ceramic pieces, metal chips, or even stainless-steel wool.

As you heat your mixture, it boils, and the vapor that comes off this liquid is *richer in the lower boiling component.* The vapor moves out of the flask and condenses, say, on the first centimeter of column packing. Now, the composition of the liquid still in the flask has changed a bit—it is *richer in the higher boiling component.* As more of *this* liquid boils, more hot vapor comes up, mixes with the first fraction, and produces a new vapor, of different composition—*richer yet in the more volatile (lower boiling) component.* And guess what? This new vapor condenses in the *second centimeter* of column packing. And again, and again, and again.

Now all these are **equilibrium steps**. It takes some time for the fractions to move up the column, get comfortable with their surroundings, meet the neighbors. . . . And if you *never* let any of the liquid–vapor mixture out of the column, a condition called **total reflux,** you might get a single pure component at the top, namely, the lower boiling, more volatile component all by itself! This is an ideal separation.

Fat lot of good that does you when you have to hand in a sample. So, you turn up the heat, let some of the vapor condense, and *take off this top fraction.* This raises hell in the column. *Nonequilibrium conditions abound—mixing. Arrgh!* No more completely pure compound. And the faster you distill, the faster you let material come over, the higher your **throughput**—the worse this gets. Soon you're at **total takeoff,** and there is no time for an equilibrium to get established. And if you're doing that, you shouldn't even bother using a column.

You must strike a compromise. Fractionally distill as slowly as you can, keeping in mind that eventually the lab does end. Slow down your fractional distillations; I've found that 5–10 drops per minute coming over into the receiving flask is usually suggested. It will take a bit of practice before you can judge the right rate for the best separation. See your instructor for advice.

15

Fractional Distillation Notes

1. Read *ALL* the notes on class 1.
2. Make *sure* you have not confused the **column** with the **condenser**. The *column is wider and has glass projections inside,* at the bottom, to hold up the packing (see Fig. 7).
3. *Don't break off the projections!*
4. Do not run water through the jacket of the column!
5. Sometimes, the column is used *without* the column packing. This is all right, too.
6. If it is necessary, and it usually is, push a wad of heavy metal wool down the column, *close to the support projections,* to support the packing chips. Sometimes the packing is made entirely of this stainless-steel wool. You can see that it is self-supporting.
7. Add the column packing. Shake the column lightly to make sure none of the packing will fall out later into your distillation.
8. With all the surface area of the packing, a lot of liquid is *held up* on it. This phenomenon is called **column holdup,** *since it refers to the material retained* in the column. Make sure you have enough compound to start with, or it will all be lost on the packing.
9. A **chaser solvent** or **pusher solvent** is sometimes used to help blast your compound off the surface of the packing material. It should have a *tremendously high boiling point relative to what you were fractionating.* After you've collected most of one fraction, some of this material is left on the column. So, you throw this chaser solvent into the distillation flask, fire it up, and start to distill the chaser solvent. As the chaser solvent comes up the column, it heats the packing material, your compound is blasted off the column packing, and more of your compound comes over. Stop collecting when the temperature starts to rise—that's the chaser solvent coming over now. As an example, you might expect *p*-xylene (B.P. 138.4°C) to be a really good chaser, or pusher, for compounds that boil lower than, say, 100°C.

But you have to watch out for the deadly azeotropes.

AZEOTROPES

Once in a while, you throw together two liquids and find that you cannot separate part of them. And I don't mean because of poor equipment or poor technique or other poor excuses. You may have an **azeotrope,** a mixture with a *constant boiling point.*

One of the best known examples is the ethyl alcohol–water azeotrope. This 96% alcohol–4% water solution will boil to dryness at *constant temperature.* It's slightly scary, since you learn that a liquid is a pure compound if it boils at a constant temperature. And you thought you had it made.

There are two types of azeotrope. If the azeotrope boils off first, it's a **minimum boiling azeotrope.** After it's all gone, if there is any other component left, only then will that component distill.

If any of the components come off first, followed by the azeotrope, you have a **maximum boiling azeotrope.**

Quiz question:
Fifty milliliters of a liquid boils at 74.8°C from the beginning of the distillation to the end. Since there is no wide boiling range, can we assume that the liquid is pure?
No. It may be a constant boiling mixture called an azeotrope.

15

You should be able to see that you have to be really careful in selecting those chaser or pusher solvents. Sure, water (B.P. 100°C) is hot enough to chase ethyl alcohol (B.P. 78.3°C) from any column packing. Unfortunately, water and ethyl alcohol form an azeotrope, and the chaser technique won't work.

CLASS 4: STEAM DISTILLATION

Mixtures of tars and oils must not dissolve in water (well, not much, anyway), so we can steam distill them. The process is pretty close to simple distillation, but you should have a way of getting *fresh hot water into the setup* without stopping the distillation.

Why steam distill? If the stuff you're going to distill is *only slightly soluble in water* and may decompose at its boiling point and the bumping will be terrible with a vacuum distillation, it is better to **steam distill.** Heating the compound in the presence of steam makes the compound boil at a lower temperature. This has to do with partial pressures of water and organic oils and such.

There are two ways of generating steam.

1. Leading steam in through an inlet tube, from a steam line, through a water trap, and thus into the system. This is classic. This is complicated. This is dangerous. You have to construct a trap to keep excess water out of the steam line, you have hot glass and rubber tubing all over, and it's pretty messy.
2. Adding hot water to the flask (Fig. 57) will generate steam and thus provide an **internal source of steam.** This method is used almost exclusively in an undergraduate organic lab for the simple reason that it is so simple.

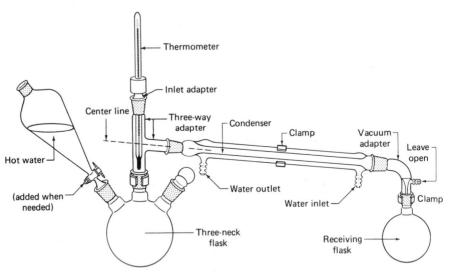

Fig. 57 An internal steam distillation setup that looks an awful lot like a simple distillation setup.

Steam Distillation Notes

1. Read *ALL* the notes on class 1 distillations.
2. Add to the distilling flask at least three times as much water (maybe more) as sample. Do not fill the flask much more than half full (three quarters, maybe). You've got to be careful. Very careful.
3. Periodically add more *hot water* as needed. When the water boils and turns to steam, it also leaves the flask, carrying product. Although exact times and quantities can't be given, here's a test.

    Collect some of the distillate, the stuff that comes over, in a small test tube. Examine the sample. If you see *two layers,* or *the solution is cloudy,* you're not done. Your product is still coming over. Keep distilling and keep adding hot water to generate more steam. *If you don't see any layers don't assume you're done.* If the sample is slightly soluble in the water, the two layers or cloudiness might not show up. Try **salting out.** This has been mentioned before in connection with **extraction** and **recrystallization** as well (see "Salting Out," p. 61, and "Extraction Hints," p. 80). Add some salt to the solution you've collected in the test tube, shake the tube to dissolve the salt, and if you're lucky, more of your product will be squeezed out of the aqueous layer, forming a *separate layer.* If that happens, *keep steam distilling* until the product does not come out when you treat a test solution with salt.

4. There should be two layers of liquid in the receiving flask at the end of the distillation. One is *mostly water.* The other is *mostly product.* To find out which is which, add a small quantity of water to the flask. The water will go into the water layer. (Makes sense.) Be very careful with this test, however; it is sometimes very hard to tell where the water has gone.
5. If you have to get more of your organic layer out of the water, you can do a **back-extraction** with an immiscible solvent (see "The Road to Recovery—Back-Extraction," p. 75).

Reflux

16

Just about 80% of the reactions in organic lab involve a step called **refluxing.** You use a reaction solvent to keep materials dissolved and at a constant temperature by boiling the solvent, condensing it, and returning it to the flask.

For example, say you have to heat a reaction to around 80°C for 17 hours. Well, you can stand there on your flat feet and watch the reaction all day. Me? I'm off to the **reflux.**

Usually, you'll be told what solvent to use, so selecting one should not be a problem. What happens more often is that you choose the reagents for your particular synthesis, put them into a solvent, and **reflux** the mixture. You boil the solvent and condense the solvent vapor *so that ALL the solvent runs back into the reaction flask* (see "Fractional Distillation," p. 113). The *reflux temperature is near the boiling point of the solvent.* To execute a reflux,

1. Place the reagents in a round-bottomed flask. The flask should be large enough to hold both the reagents and enough solvent to dissolve them, without being much more than half full.
2. You should now choose a solvent that

    a. Dissolves the reactants at the boiling temperature.
    b. Does *not* react with the reagents.
    c. Boils at a temperature that is high enough to cause the desired reaction to go at a rapid pace.

3. Dissolve the reactants in the solvent. Sometimes *the solvent itself is a reactant.* Then don't worry about 2b, above.
4. Place a condenser, upright, on the flask, connect the condenser to the water faucet, and run water through the condenser (Fig. 58). Remember—in at the bottom and out at the top.
5. Put a suitable heat source under the flask, and adjust the heat so that the solvent condenses *no higher than halfway up the condenser.* You'll have to stick around and watch for a while, since this may take some time to get started. Once the reaction is stable, though, go do something else. You'll be ahead of the game for the rest of the lab.
6. Once this is going well, leave it alone until the reaction time is up. If it's an overnight reflux, wire the water hoses on, so they don't blow off when you're not there.

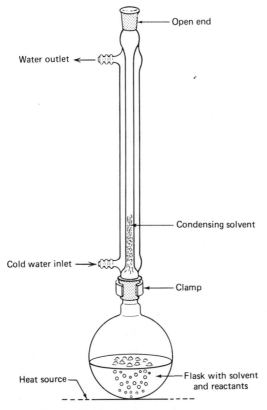

Open end

Water outlet ←

Condensing solvent

Cold water inlet →

Clamp

Heat source

Flask with solvent and reactants

Fig. 58 A reflux setup.

7.   When the reaction time is up, turn off the heat, let the setup cool, dismantle it, and collect and purify the product.

A DRY REFLUX

If you have to keep the atmospheric water vapor out of your reaction, you must use a **drying tube and the inlet adapter** in the reflux setup (Fig. 59). You can use these if you need to keep water vapor out of any system, not just the reflux setup.

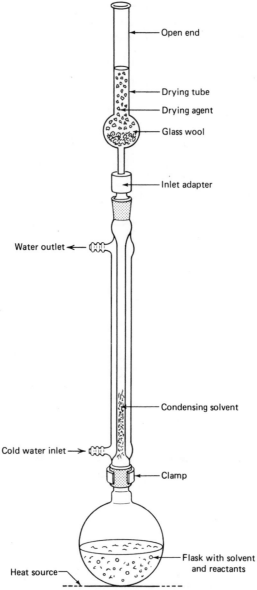

Fig. 59 Reflux setup à la drying tube.

1. If necessary, *clean and dry the drying tube.* You don't have to do a thorough cleaning unless you suspect that the anhydrous drying agent is *no longer anhydrous.* If the stuff is caked inside the tube, it is probably dead. You should clean and recharge the tube at the beginning of the semester. Be sure to use *anhydrous* calcium chloride or sulfate. It should last one semester. If you are fortunate, **indicating Drierite,** a specially prepared anhydrous calcium sulfate, might be mixed in with the white Drierite. If the color is *blue,* the drying agent is good; if *red,* the drying agent is no longer dry, and you should get rid of it (see Chapter 23, "Drying Agents").
2. Put in a loose plug of *glass wool or cotton* to keep the drying agent from falling into the reaction flask.
3. Assemble the apparatus as shown, with the *drying tube and adapter on top of the condenser.*
4. At this point, reagents may be added to the flask and heated with the apparatus. Usually, the apparatus is heated while empty to drive water off the walls of the apparatus.
5. Heat the apparatus, usually empty, on a steam bath, giving the entire setup a quarter-turn every so often to heat it evenly. A burner can be used *if there is no danger of fire* and if heating is done carefully. The heavy ground glass joints will crack if heated too much.
6. Let the apparatus cool to room temperature. As it cools, air is drawn through the drying tube before it hits the apparatus. The moisture in the air is trapped by the drying agent.
7. Quickly add the dry reagents or solvents to the reaction flask, and reassemble the system.
8. Carry out the reaction as usual, like a standard reflux.

16

ADDITION AND REFLUX

Every so often you have to add a compound to a setup while the reaction is going on, usually along with a **reflux.** Well, you *don't break open the system, let toxic fumes out, and make yourself sick to add new reagents.* You use an **addition funnel.** Now, we talked about addition funnels back with **separatory funnels** (Fig. 35) when we were considering the **stem,** and that might have been confusing.

Funnel Fun

Look at Fig. 60a. It is a true sep funnel. You put liquids in here and shake and extract them. But could you use this funnel to add material to a setup? *No.* No ground glass joint on the end; and only glass joints fit glass joints. Right? Of course, right.

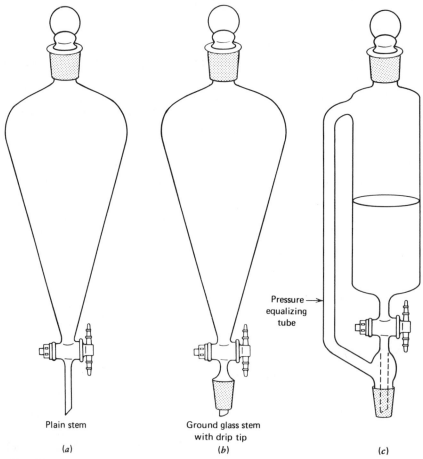

Pressure →
equalizing
tube

Plain stem   Ground glass stem
                      with drip tip

(*a*)   (*b*)   (*c*)

Fig. 60 Separatory funnels in triplicate. (*a*) Plain. (*b*) Compromise separatory / addition funnel. (c) Pressure-equalizing addition funnel.

Figure 60*c* shows a **pressure-equalizing addition funnel.** See that sidearm? Remember when you were warned to remove the stopper of a separatory funnel so you wouldn't build up a vacuum inside the funnel as you emptied it? Anyway, the sidearm equalizes the pressure on both sides of the liquid you're adding to the flask, so it'll flow freely, without vacuum buildup and without you having to remove the stopper. This equipment is very nice, very expensive, very limited, and very rare. And if you try an **extraction** in one of these, all the liquid will run out the tube onto the floor as you shake the funnel.

So a compromise was reached (Fig. 60*b*). Since you'll probably do more extractions than additions, with or without reflux, the pressure-equalizing tube went out, but the ground glass joint stayed on. Extractions; no problem. The nature of the stem is unimportant. But during additions, you'll have to take the responsibility to see that nasty vacuum buildup doesn't occur. You can remove the stopper every so often or put a **drying tube and inlet adapter** in place of the stopper. The latter keeps moisture out and prevents vacuum buildup inside the funnel.

How to Set Up

16

There are at least two ways to set up an addition and reflux, using either a **three-neck flask** or a **Claisen adapter.** I thought I'd show both these setups with **drying tubes.** They keep the moisture in the air from getting into your reaction. If you don't need them, do without them.

Often, the question comes up, "If I'm refluxing one chemical, how fast can I add the other reactant?" Try to follow your instructor's suggestions. Anyway, usually the reaction times are fixed. So I'll tell you what NOT, repeat NOT, to do.

If you reflux something, there should be a little ring of condensate, sort of a cloudy, wavy area in the barrel of the reflux condenser (Figs. 61 and 62). Assuming an exothermic reaction, the usual case, adding material from the funnel has the effect of heating up the flask. The ring of condensate begins to move *up*. Well, don't *ever* let this get more than three quarters up the condenser barrel. If the reaction is that fast, a very little extra reagent or heating will push that ring out of the condenser, and possibly into the room air. No. No, no, no.

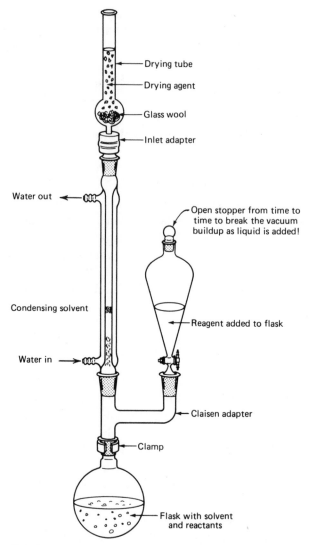

Fig. 61 Reflux and addition by Claisen tube.

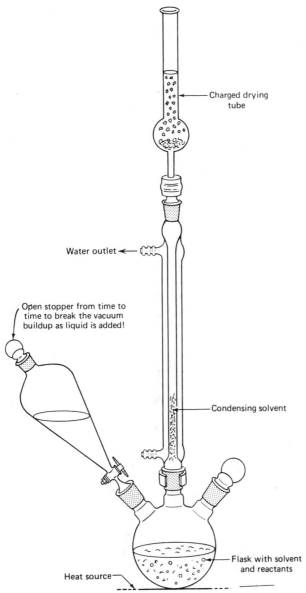

Fig. 62 Reflux and addition by three-neck flask.

Sublimation

Sublimation occurs when you heat a solid and it turns directly into a vapor. It does not pass GO nor does it turn into a liquid. If you reverse the process—cool the vapor so that it turns back into a solid—you've *condensed the vapor.* Use the unique word, **sublime,** for the direct conversion of *solid to vapor.* **Condense** can refer to either *vapor-to-solid* or *vapor-to liquid* conversions.

Figure 63 shows two forms of sublimation apparatus. Note all the similarities. Cold water goes in and down into a **cold finger** upon which

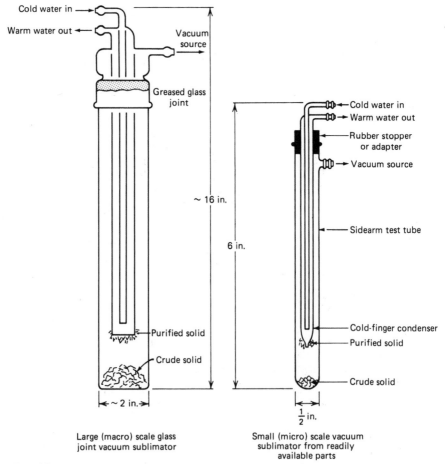

Fig. 63 King-size and miniature sublimation apparatus.

the vapors from the crystals condense. The differences are that one is larger and has a **ground glass joint.** The **sidearm test tube** with **cold-finger condenser** is much smaller. To use them,

1. Put the crude solid into the bottom of the sublimator. How much crude solid? This is rather tricky. You certainly don't want to start with so much that it touches the cold finger. And since as the purified solid condenses on the cold finger it begins to grow down to touch the crude solid, there has to be really quite a bit of room. I suggest that you see your instructor, since he may want only a small amount purified.
2. Put the cold finger into the bottom of the sublimator. Don't let the clean cold finger touch the crude solid. If you have the sublimator with the ground glass joint, lightly (and I mean lightly) grease the joint. Remember that greased glass joints should NOT be clear all the way down the joint (see Fig. 13).
3. Attach the hoses. *Cold water goes in the center tube,* pushing the warmer water out the side tube. Start the cooling water. Be careful!
4. If you're going to pull a vacuum in the sublimator, do it now. If the vacuum source is a **water aspirator,** *put a water trap between the aspirator and the sublimator.* Otherwise you may get depressed if, during a sudden pressure drop, water backs up and fills your sublimator. Also, start the vacuum *slowly.* If not, air, entrained in your solid, comes rushing out and blows the crude product all over the sublimator, like popcorn.
5. When everything has settled down, slowly begin to heat the bottom of the sublimator, if necessary. You might see vapors coming off the solid. Eventually, you'll see crystals of *purified solid form on the cold finger.* Since you'll work with different substances, different methods of heating will have to be used. Ask your instructor.
6. Now the tricky parts. You've let the sublimator cool. If you've a vacuum in the sublimator, carefully—very carefully—introduce air into the device. A sudden inrush of air, and PLOP! Your purified crystals are just so much yesterday's leftovers. Start again.
7. Now again, carefully—very carefully—remove the cold finger, with your pristine product clinging tenuously to the smooth glass surface, without a lot of bonking and shaking. Otherwise, PLOP! et cetera, et cetera, et cetera. Clean up and start again.

17

Chromatography: Some Generalities

Chromatography is perhaps the most useful means of separating compounds to purify and identify them. Indeed, separations of colored compounds on paper strips gave the technique its colorful name. Though there are many different types of chromatography, there are tremendously striking similarities among all the forms. **Thin-layer, wet-column,** and **dry-column chromatography** are common techniques you'll run across.

This chromatography works by differences in **polarity.** (That's not strictly true for all types of chromatography, but I don't have the inclination to do a 350-page dissertation on the subject, when all you might need to do is separate the differently colored inks in a black marker pen.)

ADSORBANTS

The first thing you need is an **adsorbant,** a porous material that can suck up liquids and solutions. Paper, silica gel, alumina (ultrafine aluminum oxide), corn starch, and kitty litter (unused) are all fine adsorbants. Only the first *three* are used for chromatography. You may or may not need a **solid support** with these. Paper hangs together, is fairly stiff, and can stand up by itself. Silica gel, alumina, corn starch, and kitty litter are more or less powders and will need a solid support to hold them.

Now you have an *adsorbant on some support,* or a *self-supporting adsorbant,* like a strip of paper. You also have a mixture of stuff you want to separate. So you dissolve the mixture in an easily evaporated solvent, like methylene chloride, and put some of it on the adsorbant. *Zap!* It is adsorbed! Stuck on and held to the adsorbant. But because you have a mixture of different things, and they are different, they will be *held to the adsorbant in differing degrees.*

SEPARATION OR DEVELOPMENT

Well, now there's this mixture, sitting on this adsorbant, looking at you. Now you start to *run solvents through the adsorbant.* Study the list of solvents below. Chromatographers call these solvents **eluents.**

THE ELUATROPIC SERIES

Not at all like the World Series, **the eluatropic series** is simply a list of solvents arranged according to increasing polarity.

Some solvents arranged in order of increasing polarity

(Least polar)	Pet. ether
	Cyclohexane
Increasing	Toluene
	Chloroform
Polarity	Acetone
	Ethanol
(Most polar) ↓	Methanol

So you start running "pet. ether" (i.e., petroleum ether, a mixture of hydrocarbons like gasoline—not a true ether at all). It's not very polar. So it is *not held strongly to the adsorbant.*

Well, this solvent is traveling through the adsorbant, minding its own business, when it encounters the mixture placed there earlier. It tries to kick the mixture out of the way. But most of *the mixture is more polar, held more strongly* on the adsorbant. Since the pet. ether cannot kick out compounds more polar than itself very well, most of *the mixture is left right where you put it.*

No separation.

Desperate, you try methanol, one of the most polar solvents. It is *really held strongly to the adsorbant.* So it comes along and kicks the living daylights out of just about *all* the molecules in the mixture. After all, the methyl alcohol *is more polar, so it can move right along and displace the other molecules.* And it does. So, when you evaporate the methanol and look, *all the mixture has moved with the methanol,* so you get *one spot* that moved, right with the **solvent front.**

18

No separation.

Taking a more reasonable stand, you try chloroform, because it has an intermediate polarity. The chloroform comes along, sees the mixture, and is able to push out, say, *all but one* of the components. As it travels, kicking the rest along, it gets tired and starts to leave some of the more polar components behind. After a while, *only one component is left moving with the chloroform,* and that may be dropped, too. So, at the end, *there are several spots left,* and each of them is in a *different place* from the start. Each spot is *at least one different component* of the entire mixture.

Separation. At last!

I picked these solvents for illustration. They are quite commonly used in this technique. I worry about the hazards of using chloroform, however, because it's been implicated in certain cancers. Many other common solvents, too, are suspected to be carcinogens. In lab, you will either be told what solvent (eluent) to use or you will have to find out yourself, mostly by trial and error.

Thin-Layer Chromatography

19

Thin-layer chromatography (TLC) is used for identifying compounds and determining their purity. The most common adsorbant used is **silica gel. Alumina** is gaining popularity, with good reason. Compounds should separate the same on an alumina plate as on an alumina column, and column chromatography using alumina is still very popular. And, it is very easy to run test separations on **TLC plates,** rather than carrying out tests on **chromatographic columns.**

Nonetheless, both these adsorbants are powdered and require a **solid support. Microscope slides** are extremely convenient. To keep the powder from just falling off the slides, manufacturers add a *gypsum binder* (plaster). Adsorbants with the binder usually have a "**G**" stuck on the name or say "For thin-layer use" on the container.

Sometimes a fluorescent powder is put into the adsorbant to help with **visualization** later. The powder usually glows a bright green when you expose it to **254-nm wavelength ultraviolet (UV) light.** You can probably figure out that if a container of silica gel is labeled **Silica Gel G-254,** you've got a TLC adsorbant with all the bells and whistles.

Briefly, you mix the adsorbant with water, spread the mix on the microscope slide in a *thin layer,* let it dry, then activate the coating by heating the coated slide on a hot plate. Then you **spot** or place your unknown compound on the plate, let an **eluent** run through the adsorbant (**development**), and finally examine the plate (**visualization**).

PREPARATION OF TLC PLATES

1. Clean and dry several microscope slides.
2. In an Erlenmeyer flask, weigh out some adsorbant, and add water.

    a. For silica gel use a 1:2 ratio of gel to water. About 2.5 g gel and 5 g water will do for a start.

    b. For alumina, use a 1:1 ratio of alumina to water. About 2.5 g alumina and 2.5 g water is a good start.

3. Stopper the flask, and shake it until all the powder is wet. This material MUST be used quickly because there is a gypsum (plaster) binder present.

4. Spread the mix by using a *medicine dropper. Do not use disposable pipets!* The disposable pipets have *extremely* narrow openings at the end, and they clog up easily. There exists a "dipping method" for preparing TLC slides, but since the usual solvents, methanol and chloroform *(Caution! Toxic!), do not activate the binder, the powder falls off the plate. Because the layers formed by this process are very thin, they are very fragile.*

5. *Run a bead of mix around the outside of the slide, then fill the remaining clear space. Leave ¼ in. of the slide blank on one end,* so you can hold onto the slide. *Immediately* tap the slide from the bottom to smooth the mix out (Fig. 64). Repeat this with as many slides as you can. If the mix sets up and becomes unmanageable, make up fresh adsorbant.

6. Let the slides sit until the gloss of water on the surface has gone. Then place the slides on a hot plate until they dry.

    (***CAUTION!***   *If the hot plate is too hot, the water will quickly turn to steam and blow the adsorbant off the slides.*)

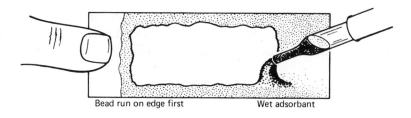

Bead run on edge first   Wet adsorbant

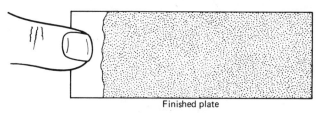

Finished plate

Fig. 64 Spreading adsorbant on a TLC plate.

19

THE PLATE SPOTTER

1. The **spotter** is the apparatus used to put the solutions you want to analyze on the plate. You use it to make a spot of sample on the plate.
2. Put the center of a **melting point capillary** atop a small, blue Bunsen burner flame. Hold it there until the tube softens and starts to sag. Do not rotate the tube, ever.
3. *Quickly* remove capillary from the flame, and pull both ends (Fig. 65). If you leave the capillary in the flame too long, you get an obscene-looking mess.
4. Break the capillary at the places shown in Fig. 65 to get **two spotters** that look roughly alike. (If you've used capillaries with *both ends open already*, then you don't have a closed end to break off.)
5. Make up 20 of these or more. You'll need them.
6. Because TLC is so sensitive, spotters tend to "remember" old samples if you reuse them. *Don't put different samples in the same spotter.*

SPOTTING THE PLATES

1. Dissolve a small portion (1–3 mg) of the substance you want to chromatograph in *any* solvent that dissolves it *and* evaporates rapidly. Dichloromethane or diethyl ether often works best.

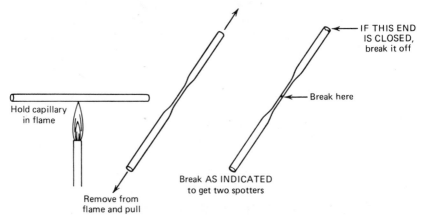

IF THIS END IS CLOSED, break it off

Break here

Hold capillary in flame

Remove from flame and pull

Break AS INDICATED to get two spotters

Fig. 65 Making capillary spotters from melting point tubes.

2. Put the thin end of the capillary spotter into the solution. The solution should rise up into the capillary.

3. Touch the capillary to the plate *briefly*! The compound will run out and form a small spot. Try to keep the spot as small as possible; NOT larger than ¼ in. in diameter. Blow gently on the spot to evaporate the solvent. Touch the capillary to the *same place*. Let this new spot grow to be *almost the same size as the one already there.* Remove the capillary and gently blow away the solvent. This will build up a concentration of the compound.

4. Take a sharp object (an old pen point, capillary tube, spatula edge, etc.) and *draw a straight line through the adsorbant,* as close to the clear glass end as possible (Fig. 66). Make sure that it runs all the way across the end of the slide and goes right down to the glass. This will keep the solvent from running up to the ragged edge of the adsorbant. It will travel only as far as the smooth line you have drawn. Measurements will be taken from this line.

5. Now make a small notch in the plate at the level of the spots to mark their starting position. You'll need this later for measurements.

DEVELOPING A PLATE

1. Take a 150-ml beaker, line the sides of it with filter paper, and cover it with a watch glass (Fig. 67).

2. Choose a solvent to **develop** the plate. You let this solvent (eluent) pass through the adsorbant by capillary action. Nonpolar eluents (solvents) will force nonpolar compounds to the top of the plate, whereas polar eluents will force *BOTH polar and nonpolar materials*

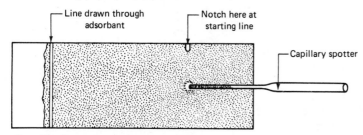

Fig. 66 Putting a spot of compound on a TLC plate.

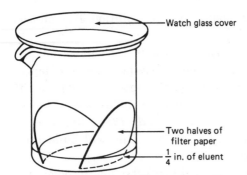

Watch glass cover

Two halves of
filter paper

$\frac{1}{4}$ in. of eluent

Fig. 67 The secret identity of a 150-ml beaker as a TLC slide development chamber is exposed.

up the plate. There is only one way to choose eluents. Educated guesswork. Use the chart of eluents in Chapter 18.

3. Pour some of the eluent (solvent) into the beaker, and tilt the beaker so that the solvent wets the filter paper. Put no more than ¼ in. of eluent in the bottom of the beaker!

4. Place the slide into the developing chamber as shown (Fig. 68). *Don't let the solvent in the beaker touch the spot on the plate* or the spot will dissolve away into the solvent! If this happens, you'll need a new plate, and you'll have to clean the developing chamber as well.

5. Cover the beaker with a watch glass. The solvent (eluent) will travel up the plate. The filter paper keeps the air in the beaker saturated with solvent so that it doesn't evaporate from the plate. When the solvent reaches the line, *immediately* remove the plate. Drain the solvent from

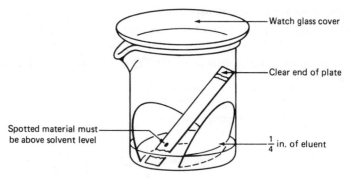

Watch glass cover

Clear end of plate

Spotted material must
be above solvent level

$\frac{1}{4}$ in. of eluent

Fig. 68 Prepared, spotted TLC plate in a prepared developing chamber.

it, and blow gently on the plate until *all* the solvent is gone. If not, there will be some trouble visualizing the spots.

Don't breathe fumes of the eluents! Make sure you have adequate ventilation. Work in a hood if possible.

VISUALIZATION

Unless the compound is colored, the plate will be blank, and you won't be able to see anything, so *you must visualize the plate.*

1. **Destructive visualization.** Spray the plate with sulfuric acid, then bake in an oven at 110°C for 15–20 min. Any spots of compound will be charred blots, utterly destroyed. *All* spots of compound will be shown.
2. **Semidestructive visualization.** Set up a developing tank (150-ml beaker) but leave out the filter paper and any solvent. Just a beaker with a cover. Add a few crystals of **iodine.** Iodine vapors will be absorbed onto most spots of compound, coloring them. Removing the plate from the chamber causes the iodine to evaporate from the plate, and *the spots will slowly disappear. Not all spots may be visible.* So if there's nothing there, that doesn't mean nothing's there. The iodine might have reacted with some spots, changing their composition. Hence the name **semidestructive visualization.**
3. **Nondestructive visualization.**
   a. **Short-wave UV (Hazard!)** Most TLC adsorbants contain a fluorescent powder that glows bright green when under short-wave UV light. There are two ways to see the spots:
      (1) The background glows green, the spots are dark.
      (2) The background glows green, the spots glow some other color. The presence of *excess eluent may cause whole sections of the plate to remain dark.* Let all the eluent evaporate from the plate.
   b. **Long-wave UV (Hazard!)** The plates stay dark. *Only the compounds may glow.* This is usually at 365 nm.

19

Both the UV tests can be done in a **UV light box,** in a matter of seconds. Since most compounds are unchanged by exposure to UV, the test is considered **nondestructive.** Not everything will show up, but the procedure is good enough for most compounds. When using the light box, *always turn it off when you leave it.* If you don't, not only does the UV filter burn out, but your instructor becomes displeased.

Since neither the UV nor the iodine test is permanent, it helps to have a record of what you've seen. You must *draw an accurate picture of the plate in your notebook.* Using a sharp-pointed object (pen point, capillary tube, etc.), you can trace the outline of the spots on the plate while they are under the UV light (*Caution! Wear gloves!*) or before the iodine fades from the plate.

INTERPRETATION

After visualization, there will be a spot or spots on the plate. Here is what you do when you look at them.

1. Measure the distance *from that solvent line* drawn across the plate *to where the spot started.*
2. Measure the distance from where the *spot stopped* to where the spot began. Measure to the *center of the spot,* rather than to one edge. If you have more than one spot, get a distance for each. If the spots are shaped funny, do your best.
3. Divide the *distance the solvent moved* into the *distance the spot(s) moved.* The resulting ratio is called the R_f **value.** *Mathematically,* the ratio for any spot should be between 0.0 and 1.0, or you goofed. *Practically,* spots with R_f values greater than about 0.8 and less than about 0.2 are hard to interpret. They could be single spots or multiple spots all bunched up and hiding behind one another.
4. Check out the R_f **value**—it may be helpful. In identical circumstances, this value would always be the *same* for a single compound, all the time. If this were true, you could identify unknowns by running a plate and looking up the R_f value. Unfortunately, the technique is not that

good, but you can use it *with some judgment* and a *reference compound* to identify unknowns (see "Multiple Spotting," below).

Figures 69 to 71 provide some illustrations. Look at Fig. 69: if you had a mixture of compounds, you could never tell. This R_f value gives no information. Run this compound again. Run a new plate. *Never redevelop an old plate!*

Use a more polar solvent!

No information in Fig. 70, either. You couldn't see a mixture if it were here. Run a new plate. *Never redevelop an old plate!*

Use a less polar solvent!

If the spot moves somewhere between the two limits (shown in Fig. 71) and *remains a single spot,* the compound is pure. If *more than one spot shows,* the compound is impure and it is a **mixture.** Whether the compound should be purified is a matter of judgment.

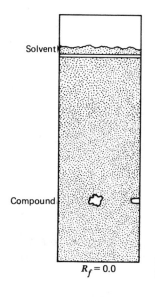

Solvent

Compound

$R_f = 0.0$

Fig. 69 Development with a nonpolar solvent and no usable results.

19

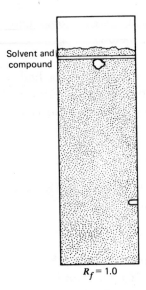

Solvent and
compound

$R_f = 1.0$

Fig. 70 Development with a very polar solvent and no usable results.

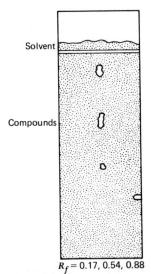

Solvent

Compounds

$R_f = 0.17, 0.54, 0.88$

Fig. 71 Development with just the right solvent is a success.

MULTIPLE SPOTTING

You can run more than one spot, either to save time or to make comparisons. You can even *identify unknowns*.

Let's say that there are two unknowns, A and B. Say one of them can be biphenyl (a colorless compound that smells like moth balls). You spot two plates. One with A and biphenyl, side by side. The other, B and biphenyl, side by side. After you develop both plates, you have the results shown in Fig. 72.

Apparently A is biphenyl.

Note that the R_f values are not perfect. This is an imperfect world, so don't panic over a slight difference.

And now we have a method that can quickly determine:

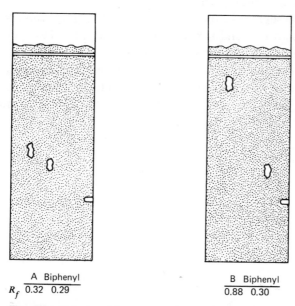

	A	Biphenyl		B	Biphenyl
R_f	0.32	0.29		0.88	0.30

Fig. 72 Side-by-side comparison of an unknown and a leading brand known.

19

1. Whether a compound is a mixture.
2. The identity of a compound *if a standard is available.*

PREPARATIVE TLC

When you use an **analytical technique** (like TLC) and you expect **to isolate compounds,** it's often called a **preparative (prep) technique.** So TLC becomes "**prep TLC.**" You use the same methods only on a larger scale.

Instead of a microscope slide, you usually use a 12×12-in. glass plate and coat it with a thick layer of adsorbant (0.5–2.0 mm). Years ago, I used a small paintbrush to put a line (a streak rather than a spot) across the plate near the bottom. Now you can get special plate streakers that give a finer line and less spreading. You put the plate in a large developing chamber and develop and visualize the plate as usual.

The thin line separates and spreads into bands of compounds, much like a tiny spot separates and spreads on the analytical TLC plates. Rather than just look at the bands though, you *scrape the adsorbant holding the different bands into different flasks,* blast your compounds off of the adsorbants with appropriate solvents, filter off the adsorbant, and finally evaporate the solvents and actually recover the separate compounds.

Wet-Column Chromatography

20

This is, as you may have guessed, chromatography carried out on a **column of adsorbant,** rather than a layer. Not only is it cheap, easy, and carried out at room temperature but you can separate large amounts, *gram quantities,* of mixtures.

In column chromatography, the adsorbant is usually alumina but can be silica gel. Except that alumina tends to be basic and silica gel, acidic, I don't know why the former is used more often. Remember, if you try out an eluent (solvent) on silica gel plates, the results on an alumina column may be different.

Now you have a *glass tube as the support* holding the adsorbant alumina in place. You dissolve your mixture and put it on the adsorbant at the top of the column. Then you wash the mixture down the column using at least one eluent (solvent), perhaps more. The compounds carried along by the solvent are washed *entirely out of the column,* into separate flasks. Then you isolate the separate fractions.

PREPARING THE COLUMN

1. The alumina is supported by a glass tube with either a stopcock or a piece of tubing and a screw clamp to control the flow of eluent (Fig. 73). You can use an ordinary buret. What you will use will depend on your own lab program. Right above this control you put a wad of cotton or glass wool to keep everything from falling out. *Do not use too much cotton or glass wool,* and *do not pack it too tightly.* If you ram the wool into the tube, the flow of eluent will be very slow, and you'll be in lab till next Christmas, waiting for the eluent. If you pack it *too loosely,* all the stuff in the column *will fall out.*

2. At this point, fill the column half-full with the least polar eluent you will use. If this is not given, you can surmise it from a quick check of separation of the mixture on a TLC plate. This would be the advantage of an alumina TLC plate.

3. Slowly put sand into the column through a funnel until there is a 1-cm layer of sand over the cotton. Adsorbant alumina is SO FINE, it is likely to go through cotton or glass wool but NOT through a layer of fine sand.

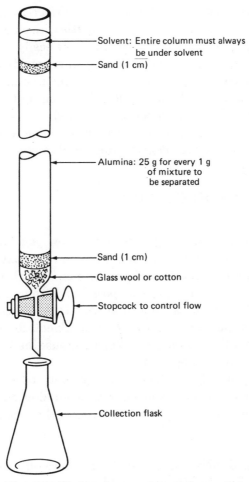

Solvent: Entire column must always be under solvent

Sand (1 cm)

Alumina: 25 g for every 1 g of mixture to be separated

Sand (1 cm)

Glass wool or cotton

Stopcock to control flow

Collection flask

Fig. 73 Wet-column chromatography setup.

4. During this entire procedure, *keep the level of the solvent above that of any solid material in the column*!
5. Now slowly add the alumina. Alumina is an adsorbant and it sucks up the solvent. When it does, heat is liberated. The solvent may boil and *ruin the column*. Add the alumina slowly! Use about 25 g of alumina for every 1 g of mixture you want to separate. While adding the

20

alumina, *tap or gently swirl the column to disloge any alumina or sand on the sides.* You know, a plastic wash bottle with eluent in it can wash the stuff down the sides of the column very easily.

6. When the alumina settles, you normally have to add sand (about 1 cm) to the top to keep the alumina from moving around.

7. Open the stopcock or clamp and let solvent out until the level of the solvent is just above the upper level of sand.

8. *Check the column! If there are air bubbles, or cracks in the column of alumina, dismantle the whole business and start over!*

COMPOUNDS ON THE COLUMN

If you've gotten this far, congratulations! Now you have to get your mixture, the analyate, on the column. Dissolve your mixture in the same solvent you are going to put through the column. Try to keep the volume of the solution of mixture as *small as possible.* If your mixture does not dissolve entirely, *and it is important that it do so,* check with your instructor! You might be able to use different solvents for the analyate and for the column, but this isn't as good. You might use the *least polar* solvent that will dissolve your compound.

If you must use the column eluent as the solvent, and not all the mixture will dissolve, you can filter the mixture through filter paper. Try to keep the volume of solution down to 10 ml or so. After this, the sample becomes unmanageable.

1. Use a pipet and rubber bulb to slowly and carefully add it to the top of the column (Fig. 74). *Do not disturb the sand!*

2. Open the stopcock or clamp and let solvent flow out until the level of the solution of compound is slightly above the sand. *At no time let the solvent level get below the top layer of sand!* The compound is now "on the column."

3. Now add eluent (solvent) to the column above the sand. *Do not disturb the sand!* Open the stopcock or clamp. Slowly let eluent run through the column until the first compound comes out. Collect the different products in Erlenmeyer flasks. You may need lots and lots of Erlen-

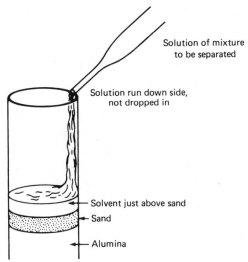

Solution of mixture
to be separated

Solution run down side,
not dropped in

Solvent just above sand

Sand

Alumina

Fig. 74 Putting compounds on the column
by pipet.

meyer flasks. *At no time let the level of the solvent get below the top of the sand!* If necessary, stop the flow, add more eluent, and start the flow again.

VISUALIZATION AND COLLECTION

If the compounds are colored, you can watch them travel down the column and separate. If one or all are colorless, you have problems. So:

1. *Occasionally* let 1 or 2 drops of eluent fall on a clean glass microscope slide. Evaporate the solvent and see if there is any sign of your crystalline compound! This is an excellent spot test, but don't be confused by nasty plasticizers from the tubing trying to put one over on you, pretending to be your product.
2. Put the narrow end of a "TLC spotter" to a drop coming off of the column. The drop will rise up into the tube. Using this loaded spotter, *spot, develop, and visualize a TLC plate with it.* Not only is this more

20

sensitive, but you can see whether the stuff coming out of the column is pure (see Chapter 19, "Thin-Layer Chromatography"). You'll probably have to collect more than one drop on a TLC plate. If it is very dilute, the plate will show nothing, even if there actually is compound there. It is best to sample 4 or 5 consecutive drops.

Once the first compound or compounds have come out of the column, those that are left may move down the column much too slowly for practical purposes. Normally you start with a nonpolar solvent. But by the time all the compounds have come off, it may be time to pick up your degree. The solvent may be too nonpolar to kick out later fractions. So you have to decide to change to a more polar solvent. This will kick the compound right out of the column.

To change solvent in the middle of a run:

1. Let old solvent level run down to just above the top of the sand.
2. *Slowly add new,* **more polar solvent** *and do not disturb the sand.*

You, and you alone, have to decide if and when to change to a more polar solvent. (Happily, sometimes you'll be told.)

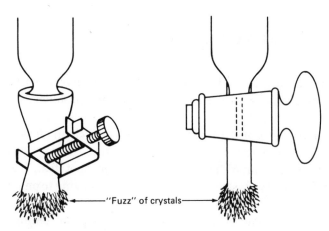

Fig. 75 A growth of crystals occurs as the eluent evaporates.

If you have only two components, start with a nonpolar solvent, and when you are sure the *first component is completely off the column, change to a really polar one.* With only two components, it doesn't matter what polarity solvent you use to get the second compound off the column.

Sometimes the solvent evaporates quickly and leaves behind a "fuzz" of crystals around the tip (Fig. 75). Just use some fresh solvent to wash them down into the collection flask.

Now all the components are off the column and in different flasks. Evaporate the solvents (*No flames!*), and lo! The crystalline material is left.

Dismantle the column. Clean up. Go home.

20

Dry-Column Chromatography

Dry-column chromatography is another approach to the separation of large quantities of a mixture of products. I think it's easier than wet-column chromatography, though more limited.

1. Weigh about 80 g of the adsorbant (alumina, silica gel, etc.) for a typical 15 × ¾-in. column, into either a large beaker or a large screw-cap bottle.

2. Get some of your eluent and *prewet the adsorbant* somewhat. For about 10 g of adsorbant, start with 3–5 g of the liquid eluent. (Oh. Did anyone ever tell you, you can weigh liquids directly, just like solids?) Now add the eluent to the adsorbant and mix like mad. You can see the advantage of a screw-cap bottle over an open beaker. The powder can't fly out of a closed bottle. *Do not add too much eluent!* You only want to precondition the adsorbant so that you don't get bubbling in the column from the heat of hydration released when you eventually run the experiment.

The powder should still flow as a powder.

3. Get a length of flexible, flat nylon tubing. Fold one end over several times (Fig. 76) and staple it to close it off.

4. Open the other end and add sand until the bottom is full (about 2–5 cm).

5. Add the *prewet, "dry"* adsorbant.

6. Cover with a layer of sand (about 1 cm).

7. Gently clamp this nylon sausage into an upright position (Fig. 76).

8. The manufacturers of dry-column chromatography adsorbants suggest piercing the bottom of the column with a few pin holes. They would know.

9. Use a disposable pipet to load your sample onto the column. The sample should go through the sand and become stuck on the adsorbant.

10. Again, with a disposable pipet, carefully add clean eluent until you're sure all the sample is stuck on the adsorbant, and none of the sand.

11. At this point, begin carefully adding more eluent. As the eluent goes down the column, the compounds in your mixture will also travel down the column at different rates, and you should get a separation.

21

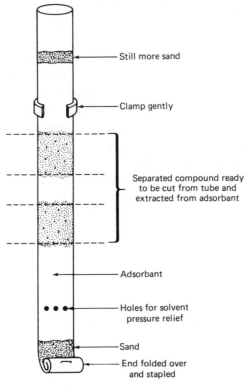

Still more sand

Clamp gently

Separated compound ready
to be cut from tube and
extracted from adsorbant

Adsorbant

Holes for solvent
pressure relief

Sand

End folded over
and stapled

DO NOT LET ELUENT RUN OUT THE END!

Fig. 76 Dry-column chromatography setup.

It is not good to keep adding eluent until it begins to drip out the
closed end onto your shoes.

12. When you're finished, you get to lie the column down carefully on
the benchtop, and slice out the sections of the column that have the
fractions you want. It's somewhat messy, but just pinch the top of the
column closed *at the level of the sand,* and drain any eluent at the top
into a waste container. Probably some sand will go with it. With *both
ends closed* nothing will move around as you move the column.

13. To recover your products, take just that section of the adsorbant with
the sample you want on it, put it into a beaker, and wash your product

off the adsorbant with a more polar solvent. Then filter off the adsorbant, strip off the solvent, and voilà! Clean, separated material.

NOTE: *Be careful with this technique for colorless samples.*
The reasons are pretty simple, if not obvious. How're you going to see where to make the cuts if you run colorless samples? Even though nylon tubing is transparent to UV light, and *theoretically* you can see the compounds under UV, many eluents absorb in the UV, and the whole column would just look dark. Then you'd probably get into trouble guessing where to make the cuts in the tubing to get the compounds separated. Just stick to colored compounds. And don't say I didn't warn you.

On
Products

The fastest way to lose points is to hand in messy samples. Lots of things can happen to foul up your product. The following are unforgivable sins!

SOLID PRODUCTS

1. ***Trash in the sample.*** Redissolve the sample, gravity filter, then evaporate the solvent.
2. ***Wet solids.*** Press out on filter paper, break up, let dry. The solid shouldn't stick to the sides of the sample vial. Tacky!
3. ***Extremely wet solids, (solid floating in water).*** Set up a gravity filtration (see p. 50), and filter the liquid off of the solid. Remove the filter paper cone with your solid product, open it up, and leave it to dry. Or remove the solid and dry it on fresh filter paper as above. Use lots of care though. You don't want filter paper fibers trapped in your solid.

LIQUID PRODUCTS

1. ***Water in the sample.*** This shows up as droplets or as a layer of water on the top or the bottom of the vial, or *the sample is cloudy.* Dry the sample with a drying agent (see Chapter 23, "Drying Agents") and gravity filter into a clean dry vial.
2. ***Trash floating in the sample.*** For that matter, it could be on the bottom, lying there. Gravity filter into a clean, dry vial.
3. ***Water in the sample when you don't have a lot of sample.*** Since solid drying agents can absorb lots of liquid, what can you do if you have a tiny amount of product to be dried? Add some solvent that has a low boiling point. It must dissolve your product. Now you have a lot of liquid to dry, and *if a little gets lost, it is not all product.* Remove this solvent after you've dried the solution. Be careful if the solvent is flammable. *No flames.*

THE SAMPLE VIAL

Sad to say, but an attractive package can sell an inferior product. So why not sell yours. Dress it up in a **neat new label.** Put on there

1. ***Your name.*** Just in case the sample gets lost on the way to camp.
2. ***Product name.*** So everyone will know what is in the vial. What does "Product from part C" mean to you? Nothing? Funny, it doesn't mean anything to instructors either.
3. ***Melting point (solids only).*** This is a range, like "M.P. 96–98°C" (see Chapter 9, "The Melting Point Experiment").
4. ***Boiling point (liquids only).*** This is a range, like "B.P. 96–98°C" (see "Distillation," p. 108). You may be asked for more data such as product yield. But the things listed above are a good start down the road to good technique.

P.S.   Gummed labels can fall off vials, and pencil will smear. *Always use ink!* And a piece of transparent tape over the label will help keep it on.

22

HOLD IT! DON'T TOUCH THAT VIAL!

Welcome to "You Bet Your Grade." The secret word is **dissolve.** Say it slowly as you watch the cap liner in some vials dissolve into your nice, clean product and turn it all goopy. A good way to prevent this is to cover the vial with aluminum foil before you put the cap on. Just make sure the product does not react with aluminum. Discuss this at length with your instructor.

Drying
Agents

23

"Dry by filtering through anhydrous sodium sulfate." This is a phrase that comes up all the time. What happens is that the anhydrous compound sucks up water and becomes hydrated. Drying agents generally leave the compound you're drying alone. Watch it though! Anhydrous calcium chloride tends to suck up alcohols. It forms an **alcohol of crystallization** as well as a **water of crystallization.** In any case, there is generally a substitute.

ANHYDROUS MAGNESIUM SULFATE

In my opinion, anh. $MgSO_4$ is about the best all-around drying agent. It has a drawback, though. Since it is a fine powder, lots of your product can become trapped on the surface. *This is not the same as water of crystallization.* The product is *only on the surface, not inside the crystal structure,* and you may wash your product off.

HOW TO DO IT

1. Put the liquid or solution to be dried into an Erlenmeyer flask.
2. Add small amounts of anhydrous magnesium sulfate until it stops caking, and you can see some of the *fine powder swirling at the bottom of the flask.*
3. Let stand a few minutes.
4. Gravity filter through filter paper (see Figs. 24–27).
5. If you've used a carrier solvent, evaporate or distill it off, whichever is appropriate. Then you'll have your clean, dry product.

FOLLOWING DIRECTIONS AND LOSING PRODUCT ANYWAY

"Add 5 g of anhydrous magnesium sulfate to dry the product." Suppose your yield of product is lower than that in the book. *Too much drying agent, not enough product—Zap!* It's all sucked onto the surface of the drying agent. Bye-bye product. Bye-bye grade.

Add the drying agent slowly to the product
in small amounts.

Now about those small amounts of product (usually liquids).

1. Dissolve your product in a *low boiling point solvent,* maybe ether or hexane or the like. Now dry this whole solution, and gravity filter. Remove, carefully, the solvent. Hoo-ha! Dried product.
2. Use chunky dehydrating agents like anhydrous calcium sulfate (Drierite). Chunky drying agents have a much smaller surface area, so not much of the product gets adsorbed.

23

DRIERITE

Drierite, one commercially available brand of anhydrous calcium sulfate, has been around a long time and is a popular drying agent. You can put it in liquids and dry them or pack a drying tube with it to keep the moisture in the air from getting into a reaction setup. But be warned. There are *two* types:

1. *White Drierite.* Ordinary anhydrous calcium sulfate chunks that you ordinarily use to dry ordinary products.
2. *Blue Drierite.* This has an indicator, a cobalt salt, that is *blue when dry, pink when wet.* Now you can easily tell when the drying agent is no good. Just look at it. Unfortunately, this stuff is not cheap, so don't fill your entire drying tube with it just because it'll look pretty. Use a small amount mixed with the white Drierite, and when the blue pieces turn pink, change the entire charge in your drying tube.

You take a chance using blue Drierite to dry a liquid directly. Sometimes the cobalt compound dissolves in your product. Then you have to clean and dry your product all over again.

Either
Ether

24

This is a short lesson in interpretation that never shows up in many lab books. It IS here because of the large number of people who have asked me time and time again,

"Is ether the same as ethyl ether?"

The answer is a resounding *yes*!

Now my ACS membership may be revoked for letting out this little secret but here goes:

Ether and ethyl ether and diethyl ether
are all the same!

Pet. ether is not the same! It means *petroleum ether,* which is *a mixture of hydrocarbons, similar to gasoline.* Needless to say, all these compounds are *very flammable* and should be handled away from all open flames, or *kaboom*!

Not Clear— Clear!

25

I've got something I want to get clear. And that is, clear is not necessarily clear. Clear? The opposite of clear, see, can be *either cloudy or colored*. To clear the confusion, stick to the term "colorless" to mean solutions that are water-white. Clear? And "cloudy" to mean solutions that, be they water-white or full of color, are nonetheless cloudy. Clear?

So solutions that are colored can be either clear or cloudy. And so too, solutions that are colorless can be either cloudy or clear.

Instrumentation in the Lab

26

Electronic instrumentation is becoming more and more common in the organic lab, which is both good and bad. The good part is that you'll be able to analyze your products, or unknowns, much faster, and potentially with more accuracy than ever. The bad part is that you have to learn about how to use the instrumentation, and there are many different manufacturers of different models of the same instrument.

The usual textbook approach is to take a piece of equipment, say something like "This is a typical model," and go on from there, trying to illustrate some very common principles. Only what if your equipment is different? Well, that's where you'll have to rely on your instructor to get you out of the woods. I'm going to pick out specific instruments as well. But at least now you won't panic if the knobs and settings on yours are not quite the same.

With that said, I'd like to point out a few things about the discussions that follow.

1. If you just submit samples to be run on various instruments, as I did as an undergraduate, pay most attention to the **sample preparation** sections. Often they say not much more than "Don't hand in a dirty sample," but often that's enough.
2. If you get to put the sample into the instruments yourself, **sample introduction** is just for you.
3. If you get to play with (in a nonpejorative sense) the instrument settings, you'll have to wade through the entire description.

You'll notice I've refrained from calling these precision instruments "machines." That's because they *are* precision instruments, not machines—unless they don't work.

Gas Chromatography

Gas chromatography (GC) can also be referred to as **vapor-phase chromatography** (VPC) and even **gas–liquid chromatography** (GLC). Usually the technique, the instrument, and the chart recording of the data are all called **GC**:

"Fire up the GC."   (the instrument)
"Analyze your sample by GC."   (perform the technique)
"Get the data off the GC."   (analyze the chromatogram)

I've mentioned the similarity of all chromatography, and just because electronic instrumentation is used, there's no need to feel that something basically different is going on.

THE MOBILE PHASE: GAS

In **column chromatography** the mobile (moving) phase is a liquid that carries your material through an adsorbant. I called this phase the **eluent,** remember? Here a gas is used to push, or *carry,* your vaporized sample, and it's called the **mobile phase.**

The **carrier gas** is usually helium, though you can use nitrogen. You use a **microliter syringe** to inject your sample into this gas stream through an **injection port,** then *onto the column.* If your sample is a mixture, the *compounds separate on the column* and reach the **detector** at different times. As each component hits the detector, the detector generates **an electric signal.** Usually the signal goes through an **attenuator network,** then out to a **chart recorder** to record the signal. I know, it's a fairly general description, and Fig. 77 is a highly simplified diagram, but there are lots of different GCs, so being specific about their operation doesn't help here. You should see your instructor. But that doesn't mean we can't talk about some things.

GC SAMPLE PREPARATION

Sample preparation for GC doesn't require much more work than handing in a sample to be graded. *Clean and dry,* right? Try to take care that the boiling point of the material is low enough to let you actually work with the technique. The maximum temperature depends on the type of column, and

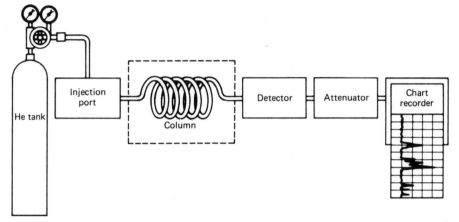

Fig. 77 Schematic of a gas chromatography setup.

that should be given. In fact, for any single experiment that uses GC, the nature of the column, the temperature, and most of the electronic settings will be fixed.

GC SAMPLE INTRODUCTION

The sample enters the GC at the **injection port** (Fig. 78). You use a microliter syringe to pierce the rubber septum and inject the sample onto the column. Don't stab yourself or anyone else with the needle. Remember, this is *not* dart night at the pub. Don't throw the syringe at the septum. There is a way to do this.

1. To load the sample, put the needle into your liquid sample and *slowly* pull the plunger to draw it up. If you move too fast, and more air than sample gets in, you'll have to push the plunger back again and draw it up once more. Usually they give you a 10-μl syringe, and 1 maybe 2 μl of sample is enough. Take the loaded syringe out of the sample, and *carefully, cautiously* pull the plunger back so there is no sample in the needle. You should see a bit of air at the very top, but not very much. This way, you don't run the risk of having your compound boil out of the needle as it enters the injector oven just before you actually inject your sample. That makes the sample broaden and reduces the resolu-

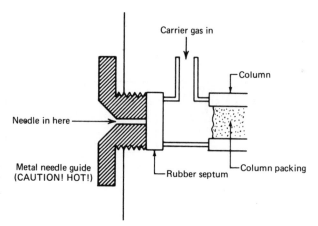

Fig. 78 A GC injection port.

tion. In addition, the air acts as an internal standard. Since air travels through the column almost as fast as the carrier gas, the **air peak** that you get can signal the start of the chromatogram, much like the notch at the start of a TLC plate. Ask your instructor.

2. Hold the syringe in *two hands* (Fig. 79). There is no reason to practice being an M.D. in the organic laboratory.

3. Bring the syringe to the level of the injection port, straight on. No angles. Then let the needle touch the septum at the center.

4. The real tricky part is holding the barrel and, without injecting, pushing the needle through the septum. This is easier to write about than it is to do the first time.

5. Now quickly and smoothly push on the plunger to inject the sample, and pull the syringe needle out of the septum and injection port.

After a while, the septum gets full of holes and begins to leak. Usually, you can tell you have a leaky septum when the pen on the chart recorder wanders about aimlessly without any sample injected.

SAMPLE IN THE COLUMN

Now that the sample is in the column, you might want to know what happens to this mixture. Did I say mixture? Sure. Just as with **thin-layer**

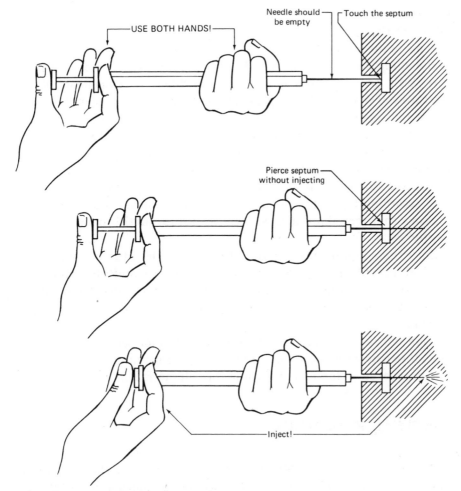

Fig. 79 Three little steps to a great GC.

27

and **column chromatography,** you can use GC to determine the composition or purity of your sample.

Let's start with two components, A and B again, and follow their path
through an **adsorption column.** Well, if A and B are different, they are
going to stick on the adsorbant to different degrees and spend more or less
time flying in the carrier gas. Eventually, one will get ahead of the other.
Aha! *Separation*—Just like column and thin-layer chromatography. Only

here the samples are vaporized, and it's called **vapor-phase chromatography (VPC).**

Some of the adsorbants are coated with a **liquid phase.** Most are very high-boiling liquids, and some look like waxes or solids at room temperature. Still, they're liquid phases. So, the different components of the mixture you've injected will spend different amounts of time in the liquid phase and, again, a separation of components in your compound. Thus the technique is known as **gas–liquid chromatography (GLC).** Thus you could use the same adsorbant, and different liquid phases, and change the characteristics of each column. Can you see how the sample components would partition themselves between the gas and liquid phases and separate according to, perhaps, molecular weight, polarity, size, and so on, making this technique also known as **liquid-partition chromatography**?

Since these liquid phases on the adsorbant are, eventually, liquids, you can boil them. And that's why there are temperature limits for columns. It is not the best to heat a column past the recommended temperature, boiling the liquid phase right off the adsorbant and right out of the instrument.

High temperature and air (oxygen) are death for some liquid phases, since they oxidize. So make sure the carrier gas is running through them at all times, even a tiny amount, while the column is hot.

SAMPLE AT THE DETECTOR

There are several types of **detectors,** devices that can tell when a sample is passing by them. They detect the presence of a sample and convert it to an electrical signal that's turned into a **GC peak** (Fig. 80) on the chart recorder. The most common type is the **thermal conductivity detector.** Sometimes called "hot-wire detectors," these devices are very similar to the filaments you find in light bulbs, and they require some care. Don't ever turn on the **filament current** unless the carrier gas is flowing. A little air (oxygen), a little heat, a little current, and you get a lot of trouble replacing the burned-out detector.

Usually there'll be at least two thermal conductivity detectors in the instrument, in a "bridge circuit." Both detectors are set in the gas stream,

Height (mm)	8	68	145
Width at half-height (mm)	4	4	4
Area (mm^2)	32	212	580
Relative area	1	8.5	18.1
Distance from injection (in.)	3	3^5/$_{16}$	3^{11}/$_{16}$
Retention times t_R at 2 min/in. (min)	6	6.63	7.38

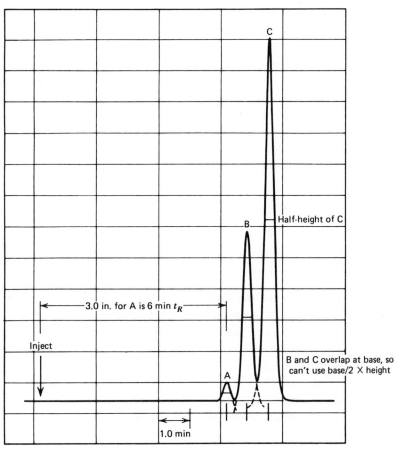

Fig. 80 A well-behaved GC trace showing a mixture of three compounds.

but only one gets to see the samples. The electric current running through them heats them up, and they lose heat to the carrier gas at the same rate.

As long as no samples, only carrier gas, goes over *both* detectors, the bridge circuit is **balanced.** There's no signal to the recorder, and the pen does not move.

Now a sample in the carrier gas goes by one detector. This sample has a thermal conductivity different from that of pure carrier gas. So the **sample detector** loses heat at a different rate from the **reference detector.** (Remember, the **reference** is the detector that NEVER sees samples—only carrier gas.) The detectors are in different surroundings. They are not really equal any more. So the bridge circuit becomes unbalanced, and a signal goes to the chart recorder, giving a GC peak.

Try to remember the pairing of **sample** with **reference** and that it's the *difference* in the two that most electronic instrumentation responds to. You will see this again and again.

ELECTRONIC INTERLUDE

There are two other stops the electrical signal makes on its way to the chart recorder.

1. *The coarse attenuator.* This control makes the signal weaker (attenuates it). Usually there's a scale marked in **powers of two:** 2, 4, 8, 16, 32, 64, ... So each position is half as sensitive as the last one. There is one setting, either an ∞ or an **S** (for **shorted**), which means that the attenuator has *shorted out the terminals connected to the chart recorder.* Now the **chart recorder zero** can be set properly.
2. *The GC Zero control.* This is a control that helps set the **zero position** on the chart recorder, but it is *not to be confused with the zero control on the chart recorder.*

Here's how to set up the electronics, properly, for a GC and a chart recorder.

1. The **chart recorder** and **GC** should be allowed to warm up and stabilize for at least 10–15 min. Some systems take more time; ask your instructor.

2. Set the **coarse attenuator** to the highest attenuation, usually an ∞ or S (for shorted).

3. Now set the pen on the chart recorder to zero using *only the chart recorder zero control*. Once you do that, *leave the chart recorder alone*.

4. Start turning the *coarse attenuator* control to more sensitive settings (lower numbers), and *watch the pen on the chart recorder*.

5. If the pen on the chart recorder moves off zero, *use the GC zero control only* to bring the pen back to the zero line on the chart recorder paper.

6. *Do not touch the chart recorder zero. Use the GC zero control only*.

7. As the *coarse attenuator* gets to more sensitive settings (lower numbers), it becomes more difficult to adjust the *chart recorder pen to zero* using *only the GC zero control*. Do the best you can at the *lowest attenuation* (highest sensitivity) you can hold a zero steady at.

8. Now, you *don't* normally run samples on the GC at attenuations of 1 or 2. These settings are *very sensitive,* and there may be lots of electrical noise—the pen jumps about. The point is, *if the GC zero is OK at an attenuation of 1,* then when you run at attenuations of 8, 16, 32, and so on, *the baseline will not jump if you change attenuation in the middle of the run*.

Now that the attenuator is set to give peaks of the proper height, you're ready to go. Just be aware that there may be a **polarity switch** that can make your peaks shift direction.

SAMPLE ON THE CHART RECORDER

27

Interpreting a GC is about the same as interpreting a TLC plate, so I'll use TLC terms as comparison to show the similarities. Remember the R_f value from TLC? The ratio of the distance the eluent moved to how far the spots of compound moved? Well, distances can be related to times, so the equivalent of R_f in GC is **retention time.** It's the time it takes the sample to move through the column minus the time it takes for the carrier gas to move through the column. Remember the part about putting air into the syringe to get an **air peak**? Well, you can assume that air travels with the carrier gas and doesn't interact with the column material. So the air peak that shows up on the chart paper can be considered to be the reference point, the "notch," as it were, that marks the start, just as on the TLC plates.

OK, so you don't want to use an air peak. Then make a mark on the chart paper as soon as you've injected the sample, and use *that* as the start. Not as good, but it'll work.

No. You do not need a stopwatch for the retention times. Find out the distance the chart paper crawls in, say, a minute. Then get out your little ruler and measure the distances from the starting point (either air peak or pen mark) to the midpoint of each peak on the baseline (Fig. 80). Don't be wise and do any funny angles. It won't help. You've got the distances and the chart speed, so you've got the retention time. It works out. Trust me.

You can also estimate how much of each compound is in your sample by measuring **peak areas.** The area under each GC peak is proportional to the amount of material that's come by the detector in that fraction. You might have to make a few assumptions (e.g., the peaks are truly triangular and each component gives the same response at the detector), but usually it's pretty straightforward. Multiply the height of the peak by the width at half the height. If this sounds suspiciously like the area of a triangle, you're on the right track. It's usually not half the base times the height, however, since sometimes the **baseline** is not very even, and that measurement is difficult.

PARAMETERS, PARAMETERS

To get the best GC trace from a given column, there are lots of things you can do, simply because there are so many controls that you have. Usually you'll be told the correct conditions, or they'll be preset on the GC.

Gas Flow Rate

The faster the carrier gas flows, the faster the compounds are pushed through the column. Because they spend *less time in the stationary phase,* they don't separate as well, and the *GC peaks come out very sharp but not well separated.* If you *slow the carrier gas down too much,* the compounds spend so much time in the stationary phase that *the peaks broaden and overlap gets very bad.* The optimum is, as always, the best separation you need, in the shortest amount of time. Sometimes the manufacturer of the

GC recommends ranges of gas flow. Sometimes you're on your own. Most of the time, someone else has already worked it out for you.

Temperature

Whether you realize it or not, the GC column has its own heater—the **column oven.** If you turn the temperature up, the compounds hotfoot it through the column very quickly. Because they spend *less time in the stationary phase,* they don't separate as well, and the *GC peaks come out very sharp but not well separated.* If you *turn the temperature down some,* the compounds spend so much time in the stationary phase that *the peaks broaden and overlap gets very bad.* The optimum is, as always, the best separation you need ᵢn the shortest amount of time. There are two absolute limits, though.

1. *Too high and you destroy the column.* The adsorbant may decompose, or the liquid phase may boil out onto the detector. *Never exceed the recommended maximum temperature for the column material.* Don't even come within 20°C of it just to be safe.
2. *Way too low a temperature, and the material condenses on the column.* You have to be above the **dew point** of the least volatile material. Not the boiling point. Water doesn't always condense on the grass—*become dew*—every day that's just below 100°C (that's 212°F, the boiling point of water). Fortunately, you don't have to know the dew points for your compounds. You *do* have to know that you *don't have to be above the boiling point* of your compounds.

27

Incidentally, the **injector** may have a separate **injector oven,** and the **detector** may have a separate **detector oven.** Set them both 10 to 20°C higher than the column temperature. You can even set these *above the boiling points* of your compounds, since you *do not want them to condense in the injection port or the detector, ever.* For those with *only one temperature control,* sorry. The injection port, column, and detector are all in the same place, all in the same oven, and all at the same temperature. The *maximum temperature,* then, is *limited by the decomposition temperature of the column.* Fortunately, because of that dew point phenomenon, you really don't have to work at the boiling points of the compounds either.

HP
Liquid
Chromatography

28

HPLC. Is it **high-performance liquid chromatography** or **high-pressure liquid chromatography** or something else? It's probably easier to consider it a delicate blend of **wet-column chromatography** and **gas chromatography** (see Chapters 20 and 27, respectively).

Rather than letting gravity **pull** the solvent through the powdered adsorbant, the liquid is **pumped** through under pressure. Initially, high pressures [1000–5000 psig (**pounds per square inch gauge**), i.e., not absolute] were used to push a liquid through a tightly packed solid. But the technique works well at lower pressures (~250 psig), hence the name high-performance liquid chromatography.

From there on, the setup (Fig. 81) resembles gas chromatography very closely. Although a **moving liquid phase** replaces the helium stream, compounds are put onto a column through an **injection port,** they *separate inside a chromatographic column* in the same way as in GC by spending more or less time in a *moving liquid* now, and the separated compounds pass *through a detector.* There the amounts of each compound as they go by the **detector** are turned into electrical signals and *displayed on a chart recorder as HPLC peaks* that look just like GC peaks. You should also get the feeling that the analysis of these HPLC traces is done in the same manner as GC traces, because it is.

Again, there's a lot of variety among HPLC systems, so what I say won't necessarily apply to your system in every respect. But it should help. I've based my observations on a Glencoe HPLC unit. It is simple and rugged, performs very well, and uses very common components carried by almost every HPLC supplier. Parts are easy to get.

THE MOBILE PHASE: LIQUID

If you use only *one liquid,* either **neat** or as a mixture, the entire chromatogram is said to be **isochratic.** There are units that can deliver *varying solvent compositions over time.* These are called **gradient elution systems.**

For an **isochratic** system, you usually use a single solution, or a neat liquid, and put it into the **solvent reservoir,** generally a glass bottle with

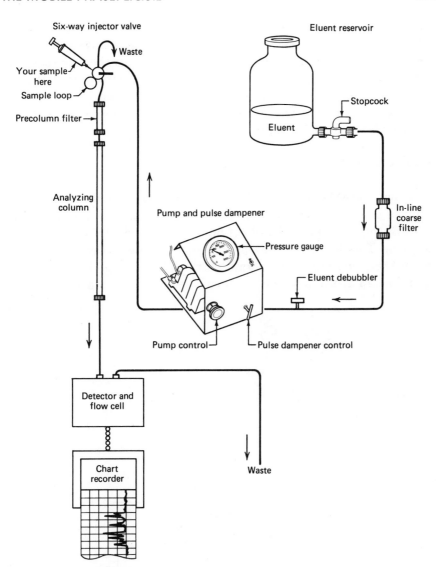

Fig. 81 Block diagram of an HPLC setup.

28

a stopcock at the bottom to let the solvent out (Fig. 82). The solvent travels out of the bottom of the reservoir and usually through a **solvent filter** that traps out any fairly large, insoluble impurities that may be in the solvent.

It is important, if you're making up the eluent yourself, to follow the directions *scrupulously*. Think about it. If you wet the entire system with the *wrong eluent,* you can wait a very long time for the *correct* eluent to reestablish the correct conditions.

A Bubble Trap

Air bubbles are the nasties in HPLC work. They cause the same type of troubles as with **wet-column chromatography,** and you just don't want them. So there's usually a **bubble trap** (Fig. 83) before the eluent reaches the pump. This device is quite simple, really. Bubbles in the eluent stream rise up the center pipe and are trapped there. To get rid of the bubbles, you open the cap at the top. Solvent then rises in the tube and pushes the

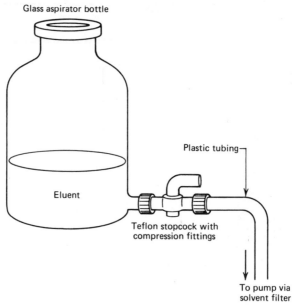

Glass aspirator bottle

Plastic tubing

Eluent

Teflon stopcock with
compression fittings

To pump via
solvent filter

Fig. 82 Aspirator bottle used to deliver eluent.

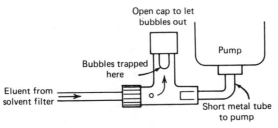

Fig. 83 Common bubble trap.

bubbles out. You have to be *extremely* careful about bubbles if you're the one to start the setup or if the solvent tank has run out. Normally, *one bubble purge per day* is enough.

The Pump

The most common pumping system is the **reciprocating pump.** Milton Roy makes a pretty good model. The pump has a **reciprocating ruby rod** that moves back and forth. On the backstroke, the pump loads up on a little bit of solvent, then it squirts it out, under pressure, on the forward stroke. If you want to increase the amount of liquid going through the system, you can dial the length of the stroke, from zero to a preset maximum, using the micrometer at the front of the pump (see Fig. 84). Use a *fully clockwise* setting, and the stroke length is zero—*no solvent flow.* A *fully counterclockwise* setting gives the maximum stroke length—*maximum solvent flow.* If you have a chance to work with this type of pump, *always turn the micrometer fully clockwise to give a zero stroke length before you start the pump.* If there *is* a stroke length set *before* you turn the pump on, the first smack can damage the reciprocating ruby rod. And it is *not* cheap.

The Pulse Dampener

Because the rod **reciprocates** (i.e., goes back and forth) you'd expect huge swings in pressure, pulses of pressure, to occur. That's why they make **pulse dampeners.** A coiled tube is hooked to the pump on the side opposite the column (Fig. 84). It is filled with the eluent that's going

28

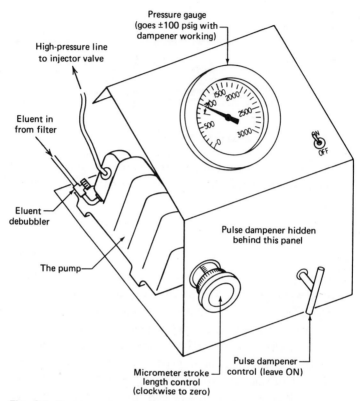

Fig. 84 Pump, pulse dampener, and pressure gauge unit.

through the system. On the forward stroke, solvent is compressed into this tube and *at the same time,* a shot of solvent is pushed onto the column. On the backstroke, while the pump chamber fills up again, the eluent we just pressed into the pulse dampener *squirts out into the column.* Valves in the pump take care of directing the flow. With the eluent in the pulse dampener tubing taking up the slack, the huge variations in pressure, from essentially zero to perhaps 1000 psig, are evened out. They don't disappear, going about 100 psig either way, but these **dampened pulses** are now too small to be picked up on the detector. They don't show up on the chart recorder either.

HPLC SAMPLE PREPARATION

Samples for HPLC must be *liquids* or *solutions*. It would be nice if the solvent you've dissolved your solid sample in were the same as the eluent.

It is *absolutely crucial* that you preclean your sample. *Any* decomposed or insoluble material will stick to the top of the column and can continually poison further runs. There are a few ways to keep your column clean.

1. ***The Swinney adapter*** (Fig. 85). This handy unit locks onto a syringe *already* filled with your sample. Then you push the sample *slowly* through a Millipore filter to trap insoluble particles. This does *not,* however, get rid of **soluble tars** that can ruin the column. (Oh. Don't confuse the filters with the papers that separate them. It's embarrassing.)

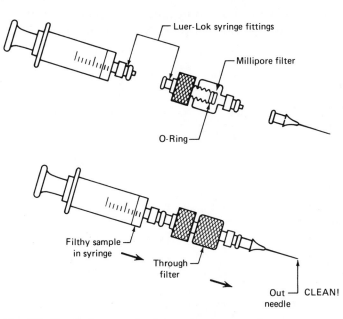

Fig. 85 The Swinney adapter and syringe parts.

28

2. ***The precolumn filter*** (Fig. 86). Add a tiny column, filled with exactly the same material as the main column, and *let this small column get contaminated*. Then unscrew it, clean out the gunk and adsorbant, and refill it with fresh column packing. The disadvantage is that you don't really know when the garbage is going to poison the entire precolumn filter and then start ruining the analyzing column. The only way to find out whether you have to clean the precolumn is to take it out of the instrument. You really want to clean it out long before the contaminants start to show up at the precolumn exit.

HPLC SAMPLE INTRODUCTION

This is equivalent of the **injection port** for the GC technique. With GC you could inject through a rubber septum directly onto the column. With HPLC it's very difficult to inject against a liquid stream moving at possibly 1000 psig. That's why they invented **injection port valves** for HPLC: you put your sample into an **injection loop** on the valve that is *not in the liquid*

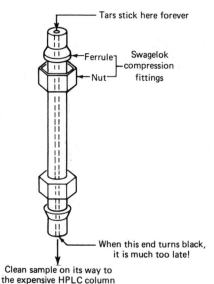

Tars stick here forever

Ferrule ⎤
          ⎬ Swagelok compression fittings
Nut ⎦

When this end turns black, it is much too late!

Clean sample on its way to the expensive HPLC column

Fig. 86   The precolumn filter.

stream, then *turn the valve,* and voilà, *your sample is in the stream,* headed for the column.

The valve (Fig. 87) has two positions.

1. ***Normal solvent flow.*** In this position, the eluent comes into the valve, goes around, and comes on out into the column without any bother. *You load the sample loop in this position.*
2. ***Sample introduction.*** Flipped this way, the eluent is *pumped through the sample loop* and any sample there is carried along and into the column. *You put the sample on the column in this position.*

SAMPLE IN THE COLUMN

Once the sample is in the column, there's not much difference between what happens here and what happens in paper, thin-layer, vapor-phase (gas), wet-column, or dry-column chromatography. *The components in the mixture will stay on the stationary phase, or move in the mobile phase for different times, and end up at different places when you stop the experiment.*

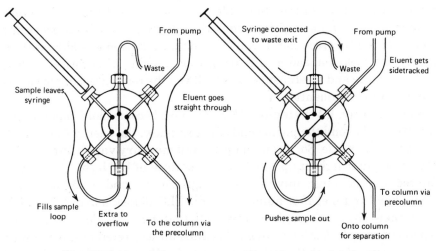

Fig. 87 The sample injector valve.

So what's the advantage? You can separate and detect microgram quantities of solid samples much as in GC. And you can't do solids all that well by GC because you have to vaporize your solid sample, probably decomposing it.

A novel development for HPLC is something called **bonded reversed-phase columns,** where the stationary phase is a nonpolar hydrocarbon, chemically bonded to a solid support. You can use these with aqueous eluents, usually alcohol–water mixtures. So you have a *polar eluent and a nonpolar stationary phase,* something that does not usually occur for ordinary wet-column chromatography. One advantage is that you don't need to use anhydrous eluents (very small amounts of water can change the character of normal phase columns) with reversed-phase columns.

SAMPLE AT THE DETECTOR

There are many HPLC detectors that can turn the presence of your compound into an electrical signal to be written on a chart recorder. Time was the **refractive index** detector was common. Clean eluent, *used as a reference,* went through one side of the detector, and the *eluent with the samples* went through the other side. A *difference in the refractive index between the sample and reference caused an electrical signal to be generated and sent to a chart recorder.* If you've read the section on **gas chromatography** and looked ahead at **infrared,** you shouldn't be surprised to find both a **sample** and a **reference.** I did tell you the reference/sample pair is common in instrumentation.

More recently, **UV detectors** (Fig. 88) have become more popular. UV radiation is beyond the purple end of the rainbow, the energy from which great tans are made. So, if you set up a **small mercury vapor lamp** with a power supply to light it up, you'll have a **source of UV light.** It's usually filtered to let through only the wavelengths of 365 and/or 254 nm. (And *where* have you met *that* number before? TLC plates, maybe?) This UV light then passes through a **flow cell** that has the *eluent and your separated sample flowing through it* against air as a reference. *When your samples come through the cell, they absorb the UV, and an electrical signal is generated.* Yes, the signal goes to a chart recorder and shows up as HPLC peaks.

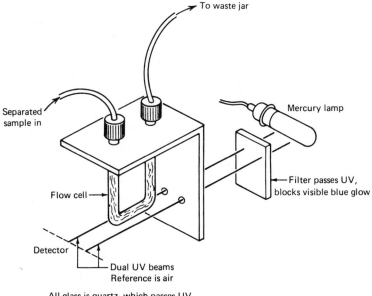

Fig. 88 Cutaway view of an HPLC UV detector.

SAMPLE ON THE CHART RECORDER

Go back and read about HPLC peak interpretation in the section on GC peak interpretation "Sample on the Chart Recorder," p. 185). The analysis is *exactly the same,* retention times, peak areas, baselines, . . . all that.

PARAMETERS, PARAMETERS

28

To get the best LC trace from a given column, there are lots of things you can do, most of them the same as for GC (see "Gas Chromatography, Parameters, Parameters," p. 186).

Eluent Flow Rate

The faster the eluent flows, the faster the compounds are pushed through the column. Because they spend *less time in the stationary phase,* they don't separate as well, and the *LC peaks come out very sharp but not well separated.* If you *slow the eluent down too much,* the compounds spend so much time in the stationary phase that *the peaks broaden and overlap gets very bad.* The optimum is, as always, the best separation you need, in the shortest amount of time. One big difference in LC is the need to worry about **back-pressure.** If you try for very high flow rates, the LC column packing tends to collapse under the pressure of the liquid. This, then, is the cause of the back-pressure, *resistance of the column packing.* If the pressures get *too high,* you may burst the tubing in the system, damaging the pump, . . . , all sorts of fun things.

Temperature

Not many LC setups have ovens for temperatues like those for GC. This is because *eluents tend to boil at temperatures much lower than the compounds on the column,* which are usually solids anyway. And eluent bubbling problems are bad enough, without actually boiling the solvent in the column. This is not to say that LC results are independent of temperature. They're not. But if a *column oven for LC is present,* its purpose more likely is *to keep stray drafts and sudden chills away* than to have a hot time.

Eluent Composition

You can vary the composition of the eluent (mobile phase) in HPLC a lot more than in GC, so there's not really much correspondence. Substitute nitrogen for helium in GC and usually the sensitivity decreases, but the retention times stay the same. Changing the mobile phases—the gases—in GC doesn't have a very big effect on the separation or retention time.

There are much better parallels to HPLC: **TLC** or **column chromatography.** Vary the eluents in these techniques and you get widely different results. With a **normal-phase silica-based column,** you can get results similar to those from silica gel TLC plates.

Infrared
Spectroscopy

Unlike the chromatographies, which physically separate materials, **infrared (IR) spectroscopy** is a method of determining what you have after you've separated it.

The **IR spectrum** is the name given to a band of frequencies between 4000 and 650 cm^{-1} beyond the red end of the visible spectrum. The units are called **wave numbers** or **reciprocal centimeters** (that's what cm^{-1} means). This range is also expressed as wavelengths from 2.5 to 15 micrometers (μm).

With your sample in the **sample beam,** the instrument scans the IR spectrum. *Specific functional groups absorb specific energies.* And because the spectrum is laid out on a piece of paper, these specific energies become *specific places* on the chart.

Look at Fig. 89. Here' a fine example of a pair of alcohols if ever there was one. See the peak (some might call it trough) at about 3400 cm^{-1} (2.9μm)? That's due to the **OH group,** specifically the stretch in the O–H bond, the **OH stretch.**

Now, consider a couple of ketones, 2-butanone and cyclohexanone (Fig. 90). There's no OH peak about 3400 cm^{-1} (2.9μm), is there? Should there be? *Of course not.* Is there an OH in 2-butanone? *Of course not.* But there is a C=O, and where's that? The peak about 1700 cm^{-1} (5.9μm). It's *not there* for the alcohols, and it *is* there for the ketones. Right. You've just correlated or *interpreted* four IRs.

Because the first two (Fig. 89) have the *characteristic OH stretch of alcohols,* they might just be alcohols. And the other two (Fig. 90) might be ketones because of the *characteristic C=O stretch* at 1700 cm^{-1} (5.9μm) in each.

What about all the other peaks? You *can* ascribe some sort of meaning to each of them, but it can be very difficult. That's why **frequency correlation diagrams,** or **IR tables,** exist (Fig. 91). They identify regions of the IR spectrum where peaks for various functional groups show up. They can get very complicated. Check to see if you can find the C—H stretch and the C—O stretch that are in all four spectra, using the correlation table. It can be fun.

For you Sherlock Holmes fans, the region from 1400 to 990 cm^{-1} (7.2–11.1 μm) is known as the **fingerprint region.** The peaks are due to the entire molecule, *its fingerprint,* rather than being from independent functional groups. And, you guessed it, no two fingerprints are alike.

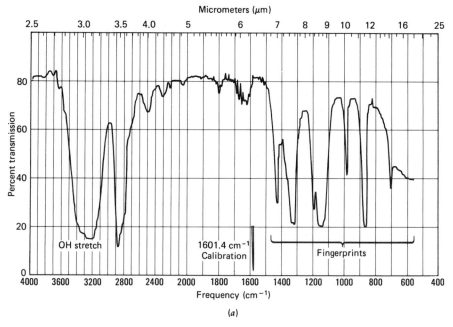

(a)

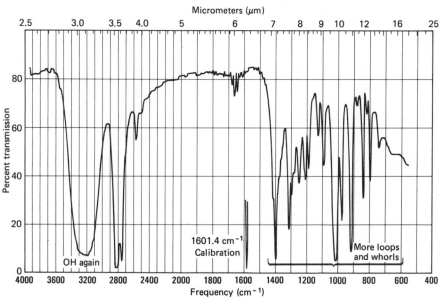

(b)

Fig. 89 IR spectra of (a) t-Butanol and (b) cyclohexanol.

29

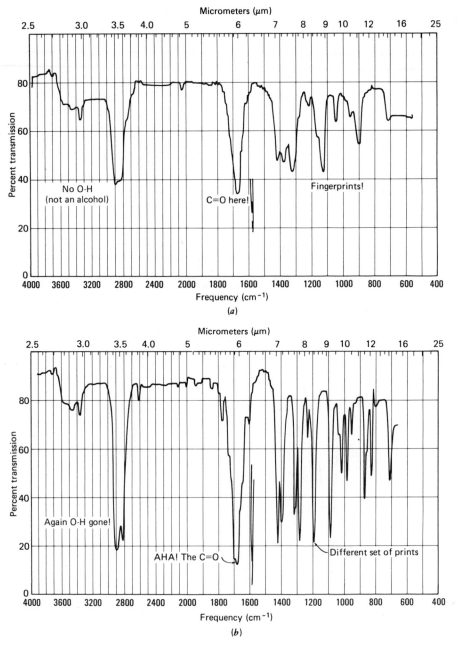

Fig. 90 IR spectra of (a) 2-butanone and (b) cyclohexanone.

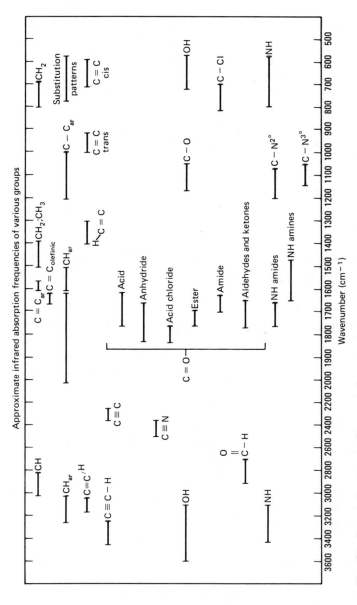

Fig. 91 The author's only IR correlation chart.

Take another look at the cyclohexanol and cyclohexanone spectra. Both have very different functional groups. Now look at the similarities, the simplicity, including the fingerprint region. Both are six-membered rings and have a high degree of symmetry. You should be able to see the similarities due to the similar structural features.

Two more things. Watch your spelling and pronunciation. It's not "infared," OK? And most people I know use IR (pronounced "eye-are," not "ear") to refer to the technique, the instrument, and the chart recording of the spectrum:

"That's a nice new IR you have there."   (the instrument)
"Take an IR of your sample."   (perform the technique)
"Let's look at your IR and see what kind of compound you have."
(interpret the resulting spectrogram)

To take an IR, you need an IR. These are fairly expensive instruments; again, no one is typical, but you can get a feeling of how to run an IR as you go on.

INFRARED SAMPLE PREPARATION

You can prepare samples for IR spectroscopy easily, but you must strictly adhere to one rule:

No water!

In case you didn't get that the first time:

No water!

Ordinarily, you put the sample between two salt plates. Yes. Common, ordinary *water-soluble* salt plates. Or mix it with **potassium bromide (K Br),** another water-soluble salt.

So keep it dry, people.

Liquid Samples

1. Make sure the sample is DRY. NO WATER!
2. Put some of the *dry* sample (2–3 drops) on one plate, then cover it with another plate (Fig. 92). The sample should spread out to cover the

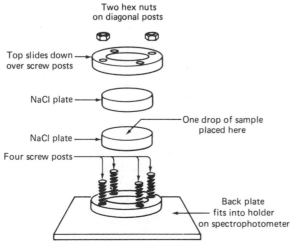

DO NOT OVERTIGHTEN

Two hex nuts
on diagonal posts

Top slides down
over screw posts

NaCl plate

One drop of sample
placed here

NaCl plate

Four screw posts

Back plate
fits into holder
on spectrophotometer

Fig. 92 IR salt plates and holder.

entire plate. You *don't have to press.* If it doesn't cover well, try turning the top plate to spread the sample, or add more sample.
3.   Place the sandwich in the IR salt plate holder, and cover it with a hold-down plate.
4.   Put at least two nuts on the posts of the holder (opposite corners) and spin them down *GENTLY* to hold the plates with an even pressure. *Do not use force!* You'll crack the plates! Remember, these are called salt plate holders and not salt plate smashers.
5.   Slide the holder and plate into the bracket on the instrument in the sample beam (closer to you, facing instrument).
6.   Run the spectrum.

Since you don't have any other solvents in there, just your liquid compound, you have just prepared a liquid sample, **neat,** meaning *no solvent.* This is the same as a *neat liquid sample,* which is a way of describing any liquid without a solvent in it. It is not to be confused with "really neat liquid sample," which is a way of expressing your true feelings about your sample.

29

Solid Samples

The Nujol Mull

A rapid, inexpensive way to get an IR of solids is to mix them with Nujol, a commercially available mineral oil. Traditionally this is called "making a Nujol mull," and it is practically idiomatic among chemists. Though you won't see Rexall or Johnson & Johnson mulls, the generic brand **mineral oil mull** is often used.

   You want to disperse the solid throughout the oil, making the solid transparent enough to IR that the sample will give a usable spectrum. Since mineral oil is a saturated hydrocarbon, it has an IR spectrum all its own. You'll find hydrocarbon bends, stretches, and push-ups in the spectrum, but you know where they are, and you ignore them. You can either look at a published reference Nujol spectrum (Fig. 93) or run your own if you're not sure where to look.

1. Put a small amount of your solid into a tiny agate mortar, and add a few drops of mineral oil.
2. Grind the oil and sample together until the solid is a fine powder *dispersed throughout the oil.*
3. Spread the mull on one salt plate, and cover it with another plate. There should be no air bubbles, just an even film of the solid in the oil.
4. Proceed as if this were a *liquid sample.*
5. Clean the plates with *anhydrous* acetone or ethanol. *NOT WATER!* If you don't have the tiny agate mortar and pestle, try a Witt spot plate and the rounded end of a thick glass rod. The spot plate is a piece of glazed porcelain with dimples in it. Use one as a tiny mortar; the other as a tiny pestle.

   And remember to forget the peaks from the Nujol itself.

Solid KBr Methods

KBr methods (hardly ever called potassium bromide methods) consist of making a mixture of your solid (*dry again*) with IR-quality KBr. Regular KBr off the shelf is likely to contain enough nitrate, as KNO_3, to give

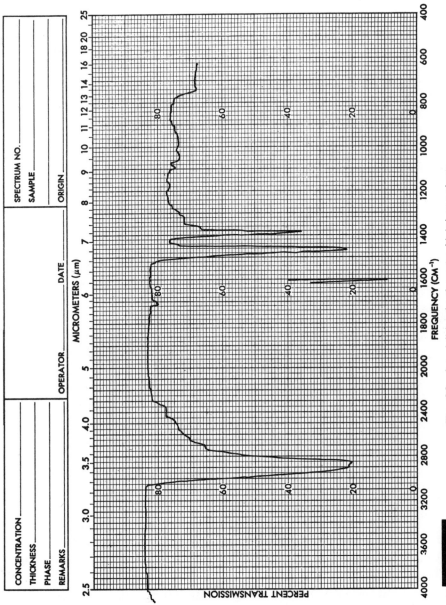

CONCENTRATION

THICKNESS

PHASE

REMARKS

SPECTRUM NO.

SAMPLE

ORIGIN

OPERATOR   DATE

Fig. 93 A published reference Nujol spectrum.

29

spurious peaks, so don't use it. After you have opened a container of KBr, dry it and later store it in an oven, with the cap off, at about 110°C to keep the moisture out.

Preparing the Solid Solution

1. At least once in your life, *weigh out* 100 mg of KBr so you'll know how much that is. If you can remember what 100 mg of KBr looks like, you won't have to weigh it out every time you need it for IR.
2. Weigh out 1–2 mg of your *dry, solid* sample. You'll have to weigh out each *sample* because different compounds take up different amounts of space.
3. Pregrind the KBr to a fine powder about the consistency of powdered sugar. Don't take forever, since moisture from the air will be coming in. Add your compound. Grind together. Serves one.

Pressing a KBr Disk—The Mini-Press (Fig. 94)

1. Get a clean, dry press and two bolts. Screw one of the bolts into the press about halfway and call that the bottom of the press.
2. Scrape a finely ground mixture of your compound (1–2 mg) and KBr (approx. 100 mg) into the press so that an even layer covers the bottom bolt.
3. Take the other bolt and turn it in from the top. *Gently* tighten and loosen this bolt at least once to spread the powder evenly on the face of the bottom bolt.
4. Hand tighten the press again, then use wrenches to tighten the bolts against each other. Don't use so much force that you turn the heads right off the bolts.
5. Remove both bolts. A KBr pellet, containing your sample, should be in the press. Transparent is excellent. Translucent will work. If the sample is opaque, you can run the IR, but I don't have much hope of your finding anything.
6. Put the entire press in a holder placed in the analyzing beam of the IR, as in Fig. 95. (Don't worry about that yet, I'll get to it in a moment. "Running the Spectrum," is next).

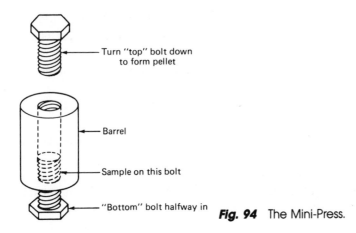

Turn "top" bolt down to form pellet

Barrel

Sample on this bolt

"Bottom" bolt halfway in

Fig. 94 The Mini-Press.

Pressing a KBr Disk—The Hydraulic Press

If you have a hydraulic press and two steel blocks available, there is another easy KBr method. It's a card trick, and at no time do my fingers leave my hands. The only real trick is you'll have to bring the card.

1. Cut and trim an index card so that it fits into the **sample beam aperture** (Fig. 100).

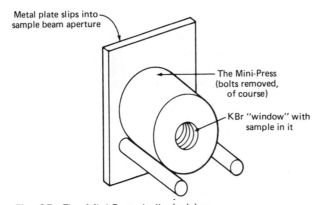

Metal plate slips into sample beam aperture

The Mini-Press (bolts removed, of course)

KBr "window" with sample in it

Fig. 95 The Mini-Press in its holder.

29

2. Punch a hole in the card with a paper punch. The hole should be *centered in the sample beam* when the card is in the sample beam aperture.

3. Place one of the two steel blocks on the bottom jaw of the press.

4. Put the card on the block.

5. Scrape your KBr–sample mixture onto the card, covering the hole and some of the card. Spread it out evenly.

6. Cover the card, which now has your sample on it, with the second metal block (Fig. 96).

7. Pump up the press to the indicated *safe* pressure.

8. Let this sit for a bit (1 min). If the pressure has dropped, bring it back up, *slowly, carefully,* to the safe pressure line, and wait another bit (1 min).

9. Release the pressure on the press. Push the jaws apart.

10. Open the metal sandwich. Inside will be a file card with a **KBr window** in it, just like in the Mini-Press.
    CAUTION! *The* KBr window *you form is rather fragile, so don't beat on it.*

11. Put the card in the **sample beam aperture** (Fig. 97). The KBr window should be *centered* in the sample beam if you've cut and punched the card correctly. Now you're ready to run the spectrum.

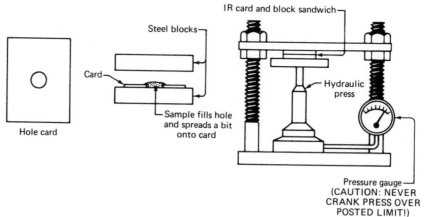

Fig. 96 KBr disks by hydraulic press.

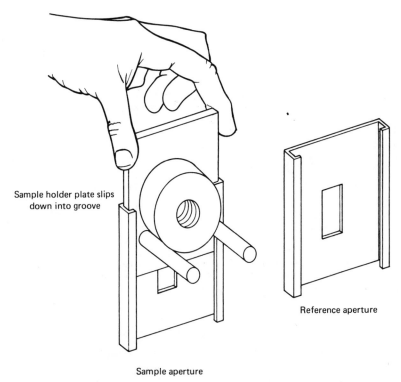

Sample holder plate slips down into groove

Reference aperture

Sample aperture

Fig. 97 Putting sample holders with samples into the beam.

RUNNING THE SPECTRUM

There are many IR instruments, and since they are so different, you need your instructor's help here more than ever. But there are a few things you have to know.

1. ***The sample beam.*** Most IRs are **dual-beam instruments** (Fig. 98). The one closest to you, if you're operating the instrument, is the **sample beam.** Logically, there is a **sample holder** for the sample beam, and your sample goes there. And there, a beam of IR radiation goes through your sample.

29

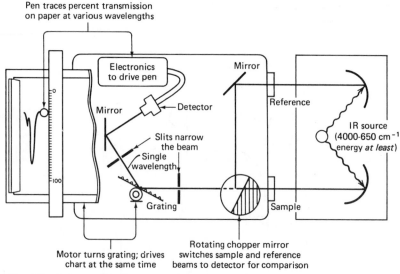

Fig. 98 Schematic diagram of an IR.

2. **The reference beam.** This is the other light path. It's not *visible* light but another part of the electromagnetic spectrum. Just remember that the reference beam is the one farthest away from you.

3. **The 100% control.** This sets the pen at the 100% line on the chart paper. Or tries to. It's a very delicate control and doesn't take kindly to excessive force. Read on and all will be made clear.

4. **The pen.** There is a pen and pen holder assembly on the instrument. This is how the spectrum gets recorded on the chart paper. Many people get the urge to throw the instrument out the window when the pen stops writing in the middle of the spectrum. Or doesn't even start writing. Or was left empty by the last fellow. Or was left to dry out on the top of the instrument. For those with Perkin–Elmer 137s or 710s, two clever fellows have made up generic felt-tip pen holders for the instruments. This way, you buy your own pen from the bookstore, and if it dries out it's your own fault. [See R.A. Bailey and J.W. Zubrick, *J. Chem. Educ.*, 59, 21 (1982).]

5. **The very fast or manual scan.** To get a good IR, you'll have to be able to scan, very rapidly, *without letting the pen write on the paper*. This is

so you'll be able to make adjustments before you commit pen to paper. This **fast forward** is not a standard thing. Sometimes you operate the instrument by hand, pushing or rotating the paper holder. Again, *do not use force.*

Whatever you do, *don't try to move the paper carrier by hand when the instrument is scanning a spectrum.* Stripped gears is a crude approximation of what happens. So, thumbs off.

I'm going to apply these things in the next section using a real instrument, the Perkin–Elmer 710B, as a model. Just because you have another model is no reason to skip this section. If you do have a different IR, try to find the similarities between it and the Perkin–Elmer model described below. Ask your instructor to explain any differences.

THE PERKIN–ELMER 710B IR (Fig. 99)

1. *On–off switch and indicator.* Press this once, the instrument comes on, and the switch lights up. Press this again, the instrument goes off, and the light goes out.
2. *Speed selector.* Selects speed (normal or fast). "Fast" is faster, but slower gives *higher resolution,* that is, more detail.
3. *Scan control.* Press this to *start a scan.* When the instrument is scanning, the optics and paper carrier move automatically, causing the IR spectrum to be drawn on the *chart paper* by the *pen.*
4. *Chart paper carriage.* This is where the chart paper nestles while you run an IR. If it looks suspiciously like a clipboard, it's because that's how it works.
5. *Chart paper hold-down clip.* Just like a clipboard, this holds the paper down in the carrier.
6. *Frequency scale.* This scale is used to help align the chart paper and to tell you during the run where in the spectrum the instrument is.
7. *Scan position indicator.* A white arrow that points, roughly, to the place in the spectrum the instrument is at.
8. *Line-up mark.* A line, here at the number 4000, that you use to

29

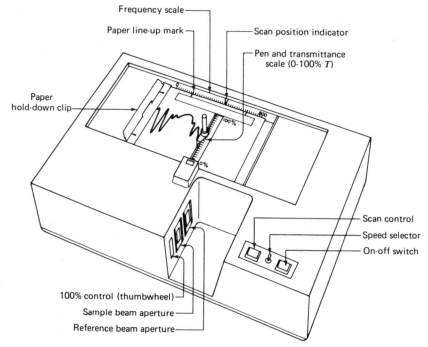

Fig. 99 The Perkin–Elmer 710B IR.

match up the numbers on the instrument frequency scale with the same numbers on the chart paper.

9. ***Pen and transmittance scale.*** This is where the pen traces your IR spectrum. The numbers here mean percentage of IR transmitted through your sample. If you have no sample in the sample beam, how much of the light is getting through? Those who said 100% are 100% correct. Block the beam with your hand and 0% gets through. You should be able to see why these figures are called the % **transmission,** or %T, scale.

10. ***The 100% control.*** Sets the pen at 100%. Or tries to. This is a fairly sensitive control, so don't force it.

11. ***Sample beam aperture.*** This is where you put the holder containing your sample, be it mull or KBr pellet. You slip the holder into the aperture window for analysis.

12. ***Reference beam aperture.*** This is where nothing goes. Or, in extreme cases, you use a **reference beam attenuator** to cut down the amount of light reaching the detector.

USING THE PERKIN–ELMER 710B

1. Turn the instrument on and let it warm up for about 3–5 min. Other instruments may take longer.
2. Get piece of IR paper and load the *chart paper carriage,* just like a clipboard. Move the paper to get the index line on the paper to line up with the index line on the instrument. It's at 4000 cm^{-1} and it's only a rough guide. Later I'll tell you how to calibrate your chart paper.
3. Make sure the chart paper carriage is at the high end of the spectrum (4000 cm^{-1}).
4. Put your sample in the sample beam. Slide the sample holder with your sample into the sample beam aperture (Fig. 97).
5. What to do next varies for particular cases. Not much, but enough to be confusing in setting things up.
6. Look at where the pen is. *Carefully* use the **100% control** to locate the pen at about the 90% mark when the chart (and spectrum) is at the high end (4000 cm^{-1}).

The 100% Control: An Important Aside

Usually there's not much more to adjusting the 100% control than is coming up in items 7 through 9, unless your sample, *by its size alone,* reduces the amount of IR reaching the detector. This really shows up if you've used the Mini-Press, which has a *much smaller opening* than that of the opening in the reference beam. So you're at a disadvantage right from the start. The 100% control mopes around at, sometimes, much less than 40%. That's terrible, and you have to use a **reference beam attenuator** (Fig. 100) to equalize the amounts of energy in the two beams. As you block more and more of the energy in the reference beam, the %T will go back to the 100% mark. Stop at about 90%. Note that this is where you'd put the pen with the 100% control anyway, if you didn't have these problems. Use the *smallest amount* of reference beam attenuation you can get away with.

29

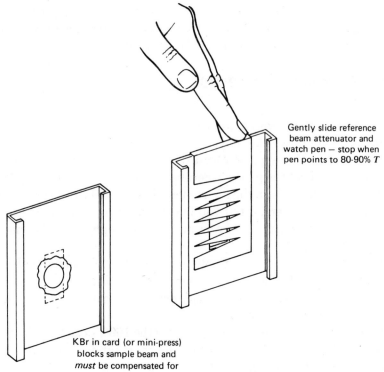

Gently slide reference
beam attenuator and
watch pen — stop when
pen points to 80-90% *T*

KBr in card (or mini-press)
blocks sample beam and
must be compensated for

Fig. 100 Using a reference beam attenuator with a KBr window.

7. OK, at 4000 cm^{-1} the %T (the pen) is at 90%.

8. Now *slowly, carefully* move the pen carriage manually so that the instrument scans the entire spectrum. *Watch the pen!* If the **baseline** creeps up and goes off the paper (Fig. 101), this is not good. Readjust the 100% control to keep the pen on the paper. Now keep going, slowly, and *if* the pen drifts up again, readjust it again with the 100% control and get the pen back on the paper.

9. Now go back to the beginning (4000 cm^{-1}). If you've adjusted the 100% control to get the pen back on the paper at some other part of the spectrum, surprise! The pen will not be at 90% when you get back. This is unimportant. What is important is that *the pen stay within the limits,* between the *0* and *100%T* lines, *for the entire spectrum.*

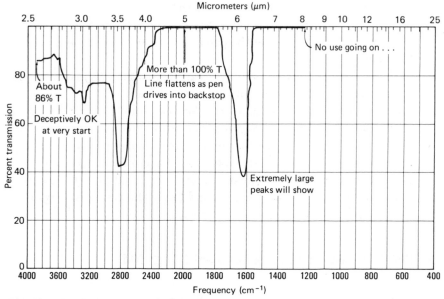

Fig. 101   An IR with an unruly baseline.

10.   In *any* case, if the peaks are too large, with the baseline in the proper place, your sample is just *too concentrated*. You can wipe some of your liquid sample or mull off one of the salt plates or remake the KBr pellet using less compound or more KBr. Sorry.

11.   When you've made all the adjustments, press the scan button, and you're off.

CALIBRATION OF THE SPECTRUM

Once the run is over, there's one other thing to do. Remember that the index mark on the paper is not exact. You have to **calibrate the paper** with a standard, usually *polystyrene film*. Some of the peaks in polystyrene are quite sharp, and many of them are very well characterized. A popular one is an extremely narrow, very sharp spike at 1601.4 cm^{-1} (6.24 μm).

1.   *Don't move the chart paper* or this calibration will be worthless.

29

2. Remove your sample, and replace it with the standard polystyrene film sample. You will have to remove any reference beam attenuator and turn the 100% control to set the pen at about 90%, when the chart is at 4000 cm^{-1}.

3. *With the pen off the paper,* move the carriage so that it's *just before* the calibration peak you want, in this case 6.24 μm.

4. Now, quickly, start the scan, *let the pen draw just the tip of the calibration peak,* and quickly stop the scan. You don't need to draw more than that (Fig. 102). Just make sure you can pick your calibration peak out of the spectral peaks. If it's too crowded at 1601.4 cm^{-1}, use a different polystyrene peak—2850.7 or 1028.0 cm^{-1} (3.51 or 9.73 μm). Anything really well known and fairly sharp will do (Fig. 103).

5. And that's it. You have a nice spectrum.

IR SPECTRA: THE FINISHING TOUCHES (Fig. 104)

On IR chart paper there are spaces for all sorts of information. It would be nice if you could fill in

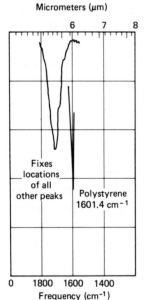

Fig. 102 A calibration peak on an IR spectrum.

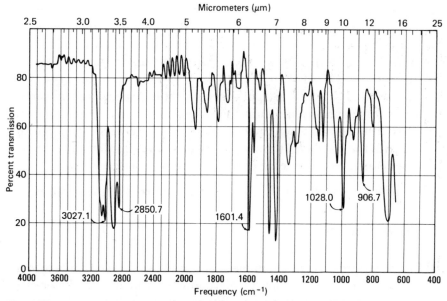

Fig. 103 IR of polystyrene film pointing out many calibration peaks.

1. **Operator.** The person who ran the spectrum. Usually you.
2. **Sample.** The name of the compound you've just run.
3. **Date.** The day you ran the sample.
4. **Phase.** For KBr, say "solid KBr." A Nujol mull is "Nujol mull."
    Liquids are either solutions in solvents or "neat liquids," that is,
    without any solvents, so call them liquids.

CONCENTRATION *Neat*		SPECTRUM NO. *6*
THICKNESS *Thin film*		SAMPLE *Cyclohexanone*
PHASE *Liquid*		
REMARKS *1601.4 cm⁻¹ Calibration*	OPERATOR *J. W. Zubrick* DATE *11/30/82*	ORIGIN *Student prep.*

Fig. 104 The finishing touches on the IR.

29

5. *Concentration.* For KBr, a solid solution, list milligrams of sample in 100 mg of KBr. For liquids, *neat* is used for liquids without solvents.
6. *Thickness.* Unless you're using solution cells, **thin film** for **neat liquids.** Leave this blank for KBr samples (unless you've measured the thickness of the KBr pellet, which you shouldn't have done).
7. *Remarks.* Tell where you put your calibration peak, where the sample came from, and anything unusual that someone in another lab might have trouble with when trying to duplicate your work. Don't put this off until the last day of the semester when you can no longer remember the details. Keep a record of the spectrum *in your notebook.*

You now have a perfect IR, suitable for framing and interpreting.

INTERPRETING IRs

IR interpretation can be as simple or as complicated as you'd like to make it. You've already seen how to distinguish alcohols from ketones by **correlation** of the positions and intensities of various peaks in your spectrum with positions listed in **IR tables** or **correlation tables.** This is a fairly standard procedure and is probably covered very well in your textbook. The things that are not in your text are

1. *Not forgetting the Nujol peaks.* Mineral oil will give huge absorptions from all the C–H bonds. They'll be the biggest peaks in the spectrum. And every so often, people mistake one of these for something that belongs with the sample.
2. *Nitpicking a spectrum.* Don't try to interpret every wiggle. There is a lot of information in an IR, but sometimes it is confusing. Think about what it is you're trying to show, then show it.
3. *Pigheadedness in interpretation.* Usually a case of, "I know what this peak is so don't confuse me with the facts." Infrared is an extremely powerful technique, but there are limitations. You don't have to go hog wild over your IR, though. I know of someone who decided that a small peak was an N–H stretch, and the compound *had* to have nitrogen in it. The facts that the intensity and position of the peak were not quite right, and neither a chemical test nor solubility studies indicated nitrogen, didn't matter. Oh well.

Nuclear Magnetic Resonance

Nuclear magnetic resonance (NMR) can be used like IR to help identify samples. But if you thought the instrumentation for IR was complicated, these NMR instruments are even worse. So I'll only give some generalities and the directions for the preparation of samples.

For organic lab, traditionally you look only at the signals from protons in your compound, so sometimes this technique is called **proton magnetic resonance (PMR).** Not naked H^+ protons either, eh? The everyday **hydrogens** in organic compounds are just called **protons** when you use this technique.

A sample in a special tube is spun between the poles of a strong magnet. A radio–frequency signal, commonly 60 MHz, a little higher than TV Channel 2, is applied to the sample. Now, were *all* protons in the same environment, there'd be this big absorption of energy in one place in the PMR spectrum. Big deal. But all protons are *not* the same. If they're closer to electronegative groups, or on aromatic rings, the signals shift to a different frequency. This change in the position of the PMR signals, which depends on the chemistry of the molecule, is called the **chemical shift.** Thus, you can tell quite a bit about a compound if you have its NMR.

LIQUID SAMPLE PREPARATION

To prepare a liquid sample for NMR analysis,

1. Get an **NMR tube.** They are about 180 mm long, 5 mm wide, and about a buck apiece for what is euphemistically called the inexpensive model. The tubes are not precision ground, and some may stick in the NMR probe. This should not be your worry, though. They also have matching, color coordinated designer caps (Fig. 105).

2. Get a **disposable pipet** and a little **rubber bulb,** and construct a **narrow medicine dropper.** Use this to transfer your sample to the NMR tube. Don't fill it much higher than about 3–4 cm. Without any solvent, this is called, of course, a **neat sample.**

3. Ask about an **internal standard.** Usually **tetramethylsilane (TMS)** is chosen because most other proton signals from any sample you might have fall at *lower frequencies* than that of the protons in TMS. Sometimes **hexamethyldisiloxane (HMDS)** is used because it

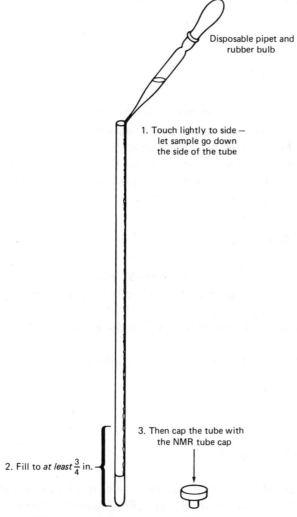

Fig. 105 Loading the typical NMR tube.

doesn't boil out of the NMR tube like TMS can. TMS boils at 26–28°C; HMDS boils at 101°C. Add *only* 1 or 2 drops.

4. Cap the tube and have the NMR of the sample taken. It's really out of place for me to tell you more about NMR here. Buy my next book, *If They Don't Work . . . They're Machines.*

30

5. Last point: cleanliness. If there is *trash* in the sample, get rid of it. Filter it or something, will you?

SOLID SAMPLES

Lucky you. You have a solid instead of a liquid. This presents one problem. What are you going to dissolve the solid in? Once it's a solution, you handle it just like a **liquid sample.** Unfortunately, if the *solvent has protons* and you know there'll be much more solvent than sample, you'll get a major proton signal from your solvent. Not a good thing, especially if the signals from your solvent and sample overlap.

Protonless Solvents

Carbon tetrachloride, a solvent without protons, is a typical protonless solvent. In fact, it's practically the only example. So if your sample dissolves in CCl_4, you're golden. Get at least 100 mg of your compound in enough solvent to fill the NMR tube to the proper height.

Caution! CCl_4 is toxic and potentially carcinogenic. Handle with *extreme care.*

Deuterated Solvents

If your compound does not happen to dissolve in CCl_4, you still have a shot because *deuterium atoms do not give PMR signals.* This is logical, since they're not protons. The problem is that *deuterated solvents are expensive,* so do NOT ask for, say, D_2O or $CDCl_3$, the deuterated analogs of water and chloroform, unless you're absolutely sure your compound will dissolve in them. Always use the protonic solvents—H_2O or $CHCl_3$ here—for the solubility test. There are other deuterated solvents, and they may or may not be available for use. Check with your instructor.

SOME NMR INTERPRETATION

I've included a spectrum of ethylbenzene (Fig. 106) to give you some idea of how to start interpreting NMRs. Obviously, you'll need more than this. See your instructor or any good organic chemistry text for more information.

The Zero Point

Look at the extreme right of the NMR. That single, sharp peak comes from the protons in the **internal standard,** TMS. This signal is *defined as zero,* and all other values for the **chemical shift** are taken from this point. The units are parts per million (ppm), and you use the Greek letter delta (δ): 0.0δ.

Protons of almost all other compounds you'll see will give signals *to the left of zero;* positive δ values, **shifted downfield** from TMS. There are compounds that give PMR signals **shifted upfield** from TMS: negative δ values.

Upfield and downfield are directions relative to where you point your finger on the NMR chart.

Signals to the *right* of where you are are **upfield.** Signals to the *left* of where you are are **downfield.**

The Chemical Shift

You can see that all the peaks don't fall in the same place, so the protons must be in different surroundings. There is one signal at 1.23δ, one at 2.75δ, and another at 7.34δ. You usually take these values from the *center* of a **split signal** (that's coming up). See that the TMS is really zero before you report the chemical shift. If it is not at zero, you'll have to add or subtract some correction to all the values. This is the same as using a polystyrene calibration peak to get an accurate fix on IR peaks.

You'll need a correlation table or a correlation chart (Fig. 107) to help in interpreting your spectrum. The $-CH_3$ group is about in the right place (1.23δ). The 7.34δ signal is from the aromatic ring, and, sure enough, that's

30

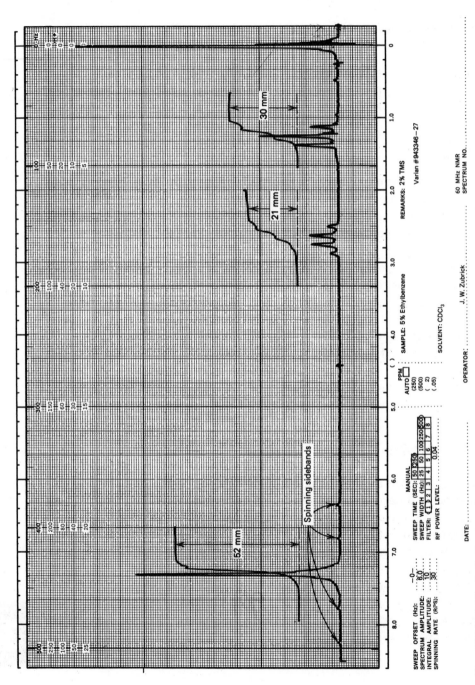

Fig. 106 NMR of ethylbenzene.

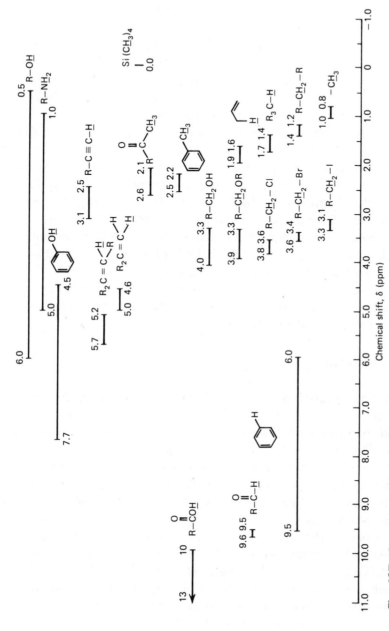

Fig. 107 A garden variety NMR correlation chart.

30

where signals from aromatic rings fall. The 2.75δ signal from the $-CH_2-$ is a bit trickier to interpret. The chart shows a $-CH_3$ on a benzene ring in this area. Don't be literal and argue that *you don't have a $-CH_3$*; you have a $-CH_2-CH_3$. All right, they're different. But the $-CH_2-$group *is on a benzene ring* and *attached to a CH_3*. That's why those CH_2 protons are further downfield; that's why you don't classify them with ordinary R–CH_2–R protons. Use some sense and judgment.

I've blocked out related groups on the correlation table. Look at the set from 3.1δ to 4.0δ. They're the areas that protons on carbons attached to halogens fall in. Read that again. It's protons *on carbons* attached to halogens. The more electronegative the halogen on the carbon, the further *downfield* the *chemical shift of those protons*. The electronegative halogen draws electrons from the carbon and thus from around the protons on the carbon. These protons, now, don't have as many electrons surrounding them. They are *not as shielded* from the big bad magnetic field as they might be. They are **deshielded,** so their signal falls *downfield*.

The hydrogen-bonded protons wander all over the lot. Where you find them, and how sharp their signals are, depends at the least on the solvent, the concentration, and the temperature.

Some Anisotropy

So what about aromatic protons $(9.5-6.0\delta)$, aldehyde protons $(9.5-9.6\delta)$, or even protons on double, nay triple bonds $(2.5-3.1\delta)$? All these protons are attached to carbons with π bonds, double or triple bonds, or aromatic systems. The electrons in these π bonds *generate their own little local magnetic field*. This local field is not *spherically symmetric*—it can shield or deshield protons depending on where the protons are—it's **anisotropic.** In Fig. 108, the **shielding regions** have plusses on them, and **deshielding regions** have minuses.

This is one of the quirks in the numbering system. Physically and psychologically, a minus means *less (less shielding)*, and DOWNfield is further left on the paper; yet the value of δ goes UP. Another system uses the Greek tau (τ)—that's $10.0 - \delta$. So 0.0δ (ppm) is 10.0τ. Don't confuse these two systems. And don't *ever* confuse deshielding (or shielding) with the *proper direction of the chemical shift*.

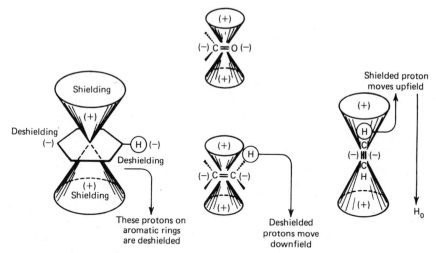

Fig. 108 Anisotropic local magnetic fields on display.

Spin–Spin Splitting

Back at ethylbenzene, you'll find that the $-CH_2-$ and the $-CH_3$ protons are not single lines. They are *split*. **Spin–spin splitting.** Such a fancy name. Protons have a spin of plus or minus ½. If I'm sitting on the methyl group, I can see two protons on the adjacent carbon ($-CH_2-$). (**Adjacent carbon,** remember that.) They spin, so they produce a magnetic field. Which way do they spin? That's the crucial point. Both can spin one way, *plus*. Both can spin one way, *minus*. Or, each can go a different way; one plus, one minus.

Over at the methyl group (**adjacent carbon,** eh?), you can feel these fields. They add a little, they subtract a little, they cancel a little. So your methyl group splits into three peaks! It's split *by* the *two* protons *on the adjacent carbon.*

Don't confuse this with the fact that there are three protons on the methyl group! THAT HAS NOTHING TO DO WITH IT! It is mere coincidence.

The methyl group shows up as a **triplet** because it is split *BY TWO* protons *on the ADJACENT carbon.*

30

Now what about the intensities? Why's the middle peak larger? Get out a marker and draw an A on one proton and a B on the other. OK. There's only one way for A and B to spin in the same direction—*Both A and B are plus* or *both A and B are minus*. But there are *two ways* for them to spin opposite each other—*A plus with B minus; B plus with A minus.* This condition happens *two times.* Both A and B plus happen only *one time.* Both A and B minus happen only *one time.* So what? So the **ratio of the intensities is 1:2:1.** Ha! You got it—a **triplet.** Do this whole business sitting on the $-CH_2-$ group. You get a **quartet**—*four lines—because the $-CH_2-$ protons are adjacent to a methyl group.* They are *split BY three to give FOUR lines* (Fig. 109).

No, that is not all. You can tell that the $-CH_2-$ protons and the $-CH_3$ protons split each other by their **coupling constant,** the distance between the split peaks of a single group. Coupling constants are called **J values,** and usually are given in hertz (Hz). You can read them right from the chart, which has a grid calibrated in hertz. If you find protons at different

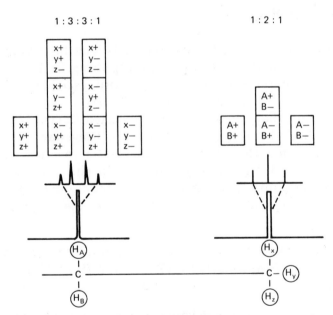

Fig. 109   Spin alignments for the ethyl group.

chemical shifts, and their coupling constants are the same, they're splitting and coupling with each other.

Integration

Have you wondered about those funny curves drawn over the NMR peaks? They're **electronic integrations,** and they can tell you how many protons there are at each chemical shift. Measure the distances between the horizontal lines just before and just after each group. With a cheap plastic ruler I get 52 mm for the benzene ring protons, 21 mm for the CH_2 protons, and about 30 mm for the CH_3 protons. Now you divide all the values by the smallest one. Well, 21 mm is the smallest, and without a calculator I get $2.47:1:1.43$. Not even close. And how do you get that 0.47 or 0.43 proton? Try for the simplest *whole number ratio*. Multiply everything by 2, and you'll have $4.94:2:2.86$. This is very close to $5:2:3$, the actual number of protons in ethylbenzene. Use other whole numbers; the results are not as good and you can't justify the splitting pattern—*3 split BY 2* and *2 split BY 3*—with other ratios. Don't use each piece of information in a vacuum.

There are a lot of other things in a typical NMR. There are **spinning sidebands,** small duplicates of stronger peaks, evenly spaced from the parent peak. They fall at *multiples of the spin rate,* here about 30 Hz. Spin the sample tube faster and these sidebands move farther away; slow the tube and they must get closer.

Signals that split each other tend to lean toward each other. It's really noticeable in the triplet and even distorts the intensity ratio in the quartet some. Ask your instructor or see another textbook if you have questions.

30

Index